COLOR ATLAS OF MUSCLE HISTOCHEMISTRY

Roger A. Brumback, MD
Formerly
Associate Professor of Neurology
University of North Dakota School of Medicine
Fargo, North Dakota

Richard W. Leech, MD
Neuropathologist
The Neuropsychiatric Institute
and
Professor of Neuropathology
University of North Dakota School of Medicine
Fargo, North Dakota

PSG PUBLISHING COMPANY, INC.
Littleton, Massachusetts

Library of Congress Cataloging in Publication Data

Main entry under title:

Brumback, Roger A.
Color atlas of muscle histochemistry.

Bibliography: p.
Includes index.
1. Muscle—Atlases. 2. Histochemistry—Atlases.
I. Leech, Richard W. II. Title. [DNLM: 1. Muscles—Anatomy and histology—Atlases. WE 17 B893c]
QM571.B78 1984 611'.0186 84-5349
ISBN 0-88416-493-4

Published by:
PSG PUBLISHING COMPANY, INC.
545 Great Road
Littleton, Massachusetts 01460

Printed in the United States of America

International Standard Book Number: 0-88416-493-4

Library of Congress Catalog Card Number: 84-5349

This book is dedicated to
W. King Engel, MD, a pioneer
in the histological study of muscle diseases.

ABOUT THE AUTHORS

Roger A. Brumback, MD, is a pediatric neurologist who received his pediatric training at the Johns Hopkins Hospital, his neurology training at the Washington University (St. Louis) Medical Center, and his training in neuromuscular disease at the National Institutes of Health under Dr. W. King Engel. Dr. Brumback, while on the faculty of the Department of Neuroscience of the University of North Dakota School of Medicine in Fargo, North Dakota, was the Director of the Red River Valley Chapter Muscular Dystrophy Association Clinic. In addition to his studies of muscle histochemistry, Dr. Brumback's research efforts have included development of an animal model of McArdle's disease and the evaluation of ion current changes in myotonia. Dr. Brumback has also published a number of articles on electroconvulsive therapy, learning disabilities, mania, depression, and the psychosocial impact of myotonic dystrophy and Duchenne dystrophy. Dr. Brumback is the coauthor of a textbook of neurology (*Practical Neurology for the Primary Care Physician*, C.C. Thomas, 1981) and recently coedited *The Neuromuscular Junction* (Futura, 1984). Currently, Dr. Brumback is a member of the Department of Pathology of the University of Rochester Medical Center, Rochester, New York.

Richard W. Leech, MD, is Director of the Neuropathology Laboratory of The Neuropsychiatric Institute Hospital, Fargo, North Dakota, and Professor and Chairman of the Division of Neuropathology of the Department of Neurosciences of the University of North Dakota School of Medicine. His responsibilities include the diagnostic neuropathology for both North and South Dakota, the latter through the auspices of the University of South Dakota School of Medicine and the Laboratory for Clinical Medicine, Sioux Falls, South Dakota. He has previously coauthored two neuropathology textbooks. The first, *Normal and Abnormal Development of the Human Nervous System* (Harper & Row, 1975) was prepared while he was at the University of Washington School of Medicine, and the second, *Neuropathology—A Summary for Students* (Harper & Row, 1982), was coauthored with Dr. Robert Shuman. He is currently working on a book on hydrocephalus, designed for paramedical personnel.

CONTENTS

FOREWORD

The histological study of skeletal muscle has become increasingly important for neurologists, rheumatologists, and orthopedists. Too often, a small sample of muscle in a jar of formalin is sent to the pathologist for interpretation. Little useful pathological information is obtained, however, from microscopic examination of such a formalin-fixed, paraffin-embedded muscle specimen. For the clinician to obtain the necessary diagnostic information from a muscle biopsy specimen, the muscle sample must be carefully removed by the surgeon, then taken in the fresh state to the pathology laboratory, where the specimen is quick-frozen. Later, sections are cut from the frozen blocks and stained, using the appropriate histochemical staining techniques.

The following case illustrates the need for histochemical techniques in muscle interpretation:

> A 47-year-old man presented for evaluation of elevated serum creatine phosphokinase (ten times normal), intermittent fever, and progressive proximal muscle weakness of several months duration. Two muscle biopsy specimens were obtained. One was placed in formalin for paraffin-embedded sections; the other was sent for fresh quick-frozen histochemical examination. The specimens were interpreted independently by the pathologists, who were aware of the clinical diagnosis of polymyositis.

The report of the formalin-fixed, paraffin-embedded sections read:

> *Macroscopic Examination*: The specimen is marked left quadriceps muscle biopsy. It is a piece of muscle 1 × 0.5 × 0.5 cm. It is rubbery and totally processed.
> *Microscopic Examination*: Section shows striated muscle with intact microscopic anatomical structure. It includes not only the muscle fibers but the vascular system and connective tissue and other related elements.
> *Pathologic Diagnosis*: Biopsy of striated muscle.

In contrast, the report of the fresh, quick-frozen histochemical sections read:

> *Gross Description*: Specimen consists of fragments of muscle submitted for histochemistry.
> *Microscopic Description*: Examination of the H&E and trichrome stains shows that there is marked fiber size variation, affecting both fiber types. There are numerous small foci of chronic inflammatory cells within the interstitial areas. The inflammatory cells are composed of lymphocytes predominantly with small foci containing plasma cells. The blood vessels appear unaffected. There are a moderate number of fibers undergoing necrosis, as well as numerous basophilic fibers containing vesicular nuclei. Other fibers show an increase in the number of internal nuclei. With the

ATPase stain no specific abnormalities are noted. There is no evidence of fiber type grouping or predominance. There is no evidence of perifascicular atrophy. With the NADH stain the intermyofibrillar network is irregular in many fibers. In addition, there are many fibers showing either decreased or increased central staining. With the alkaline phosphatase stain there is increased staining in many of the interstitial areas.

Pathologic Diagnosis: Chronic focal interstitial inflammation, marked fiber size variation and presence of necrotic and basophilic fibers which are the alterations characteristic of an inflammatory myopathy such as polymyositis.

It is evident from examining these two reports that the report of the formalin-fixed specimen would have been of little value to the clinician, while the report of the histochemically examined specimen confirmed the clinical impression of a treatable inflammatory myopathy.

Although muscle diseases are relatively rare (incidence probably one or two per 1000 population), proper diagnosis can best be made with the use of histochemical techniques. This approach is emphasized by Victor Dubowitz who, in a 1975 review article in the *Archives of Disease in Childhood* (1975;50: 335–346) makes the bold statement:

> In summary, what has been gained from the modern techniques of investigation of neuromuscular problems is the recognition of a number of different disorders which can present in a clinical fashion identical to the dystrophies or the neurogenic syndromes, and which would not be diagnosed other than by histochemical . . . techniques. I think it is important for all children to have a muscle biopsy, even in the face of an apparently clear-cut clinical diagnosis of muscular dystrophy, or on the other hand in the face of a normal EMG and serum creatine phosphokinase level. It is also essential that the biopsy, which can be done under local anaesthetic, be adequately investigated by modern techniques. It is probably unethical now to subject a child to a muscle biopsy and then to drop the specimen into a pot of formalin and essentially destroy it for any further meaningful investigation.

To provide the best possible diagnostic capabilities, we believe it is imperative that all anatomical pathology laboratories become familiar with muscle histochemical techniques. Somehow the opinion has pervaded the pathology community that such processing and staining techniques are too difficult, that interpretation of the sections is too complicated, and that the cost of muscle histochemistry is prohibitive. Actually, with a little effort a quality surgical specimen can be easily obtained, and the freezing and sectioning techniques are sufficiently similar to those used in many laboratories for other frozen sections of surgical specimens. We have produced this atlas so that the techniques of microscopic evaluation of muscle specimens can be explained.

The cost of histochemical study is not prohibitive. Dewar flasks for the liquid nitrogen used to freeze the specimen would need to be purchased at a one-time expense of approximately $120. The largest ongoing cost to the laboratory would be technician time of 3 hours for a battery of five stains (modified Gomori trichrome, H&E, alkaline myofibrillar ATPase, NADH-TR, and nonspecific esterase), with more hours necessary if further stains are required. Specimens can be mounted on an inexpensive section of 1-inch wood doweling, or plastic embedding blocks can be purchased at approximately $.75 each. Other equipment, such as an analytical balance, small incubator, pH meter, −80°C freezer, long forceps, and metal beaker can be readily purchased. Since the sections are cut onto #2 thickness coverslips, Columbia staining jars and coverslip staining racks would need to be purchased at a one-time cost of approximately $150. A Tissue-Tek cryostat, available in most laboratories, is preferred for cutting the tissue sections. Approximately 3 L of liquid nitrogen, ordered from local suppliers at approximately $2 per liter, is necessary to freeze each biopsy and can be saved and reused. Many of the chemicals necessary for the staining procedures (see Appendix) are readily available in surgical pathology laboratories, and special dyes and reagents not usually available can be purchased from large biological supply companies (such as Sigma Chemical Co). The actual cost per test for most individual histochemical stains is approximately $1.00. Certainly these expenses are reasonable, considering that most laboratories charge from $250 to $400 for this type of pathology report.

We hope this text encourages pathologists to consider establishing these techniques in their laboratory.

ACKNOWLEDGMENTS

The authors thank David Roberts, photographer, and Gary Baune, illustrator, for their help with this book. Our histotechnologists, Lilia Lopez and Margo Iszler, with their interest, skill, and attention to detail, made this atlas possible. We also thank our typist, Sandra Amundson, who was able to endure our many corrections to produce the manuscript. The Neuropsychiatric Institute, Fargo, North Dakota provided financial support for this project.

1 MUSCLE BIOPSY TECHNIQUE

The technique of muscle biopsy is relatively simple, and with a little care satisfactory tissue specimens can easily be obtained. Basically, there are two satisfactory techniques of muscle biopsy: open biopsy and needle biopsy. Open biopsy is preferable in most situations, since it provides enough tissue for adequate study. The needle biopsy does have the advantage of avoiding a large surgical incision; however, in some cases the small sample size may provide too few muscle fibers for study.

One of the most important aspects of muscle biopsy, selection of the biopsy site, is usually neglected. Too often the physician calls upon the surgeon to perform a muscle biopsy and with no further instruction the surgeon unknowingly selects an inappropriate muscle for the biopsy. In earlier times, an electromyogram (EMG) was done in multiple muscles and the muscle with the greatest abnormality was selected for biopsy. However, this technique led to many erroneous diagnoses, since needle insertion into a muscle damages fibers and produces inflammatory cell infiltration, which can be mistaken for an inflammatory myopathy. There are a number of considerations in deciding upon the correct muscle for biopsy.

First, it is important that the muscle specimen to be studied has no needle punctures or injections from the previous four months. In addition, muscle that has previously had multiple injections of sclerosing-type drugs (such as penicillin, pentazocine, phenothiazines) should *not* be biopsied. Medications and vaccines are most commonly injected into deltoid and gluteal muscles; therefore, these muscles should not be biopsied. Occasionally, quadriceps muscles are used for injections; if so, these muscles too should be avoided.

Second, in selecting the muscle for biopsy the clinician must consider muscle strength. The MRC scale of grading muscle strength, from Grade 0 to Grade 5, is as follows:

Grade 5	Normal strength
Grade 4	Movement against resistance
Grade 3	Movement against gravity
Grade 2	Movement with gravity effect eliminated
Grade 1	Flicker movement
Grade 0	No movement

In chronic muscle problems a biopsy of a muscle with strength Grade 3 or weaker will usually show severe changes, suggesting end-stage muscle; such changes cannot be evaluated as to the original cause because of the marked replacement of muscle with connective tissue and fat. In chronic muscle diseases, it is necessary that a muscle with Grade 4 or greater strength be biopsied in order to diagnose the disorder. This is not necessarily true in acute muscle diseases, in which muscles of weaker grades show the most diagnostic histological findings.

Third, there are some technical problems that the surgeon must overcome. Damage to major nerves is possible during the biopsy procedure. For example, the median nerve can be damaged during biopsy of the biceps muscle, the sciatic nerve can be damaged in biopsy of the hamstrings muscles, and the peroneal or tibial nerves can be damaged in biopsy of the gastrocnemius muscle.

Finally, the surgeon must be aware of the cosmetic aspects of open-biopsy scarring. For example, a scar over the biceps muscle would be noticeable for many patients wearing short-sleeve clothing; a scar over the gastrocnemius would be visible for patients wearing clothing which does not cover that area; a scar in the upper thigh over the quadriceps would be hidden by most clothing. Often, however, in an attempt to minimize scarring, an incision is made so small that an adequate sample of muscle for study is not obtained, necessitating a second biopsy. In performing a muscle biopsy, it is imperative that adequate muscle samples be removed from a moderately affected muscle, and from a muscle which has not recently been traumatized by intramuscular needles or injections.

An open muscle biopsy generally can be performed using local anesthesia. Since malignant hyperthermia (triggered by halogenated general anesthetics and depolarizing muscle relaxants) is more common in patients with neuromuscular diseases, local anesthesia is preferable. For most adults a local anesthetic is sufficient. For particularly anxious patients, a mild anxiolytic-sedative (such as a benzodiazepine) may be used in conjunction with the local anesthetic. In older children, greater sedation may be necessary, in which case intravenous barbiturate or inhalation nitrous oxide is used with the local anesthetic. In infants or young children, ketamine or Innovar may be necessary for adequate sedation.

OPEN MUSCLE BIOPSY

Because of the risk of infection, a sterile technique must be used in performing a muscle biopsy (Figure 1). After appropriate cleansing and draping, local anesthetic is injected subcutaneously to deaden the skin. We prefer the use of a 2% solution of lidocaine with epinephrine for adults and a 1% solution of lidocaine with epinephrine for children. About 20 to 30 mL of local anesthetic should be infiltrated subcutaneously in adults, 10 to 20 mL in children. The local anesthetic must be injected just below the skin and *not into the muscle*. The injection of local anesthetic into the muscle will damage the muscle and make interpretation of the subsequent biopsy specimens difficult. In order to avoid injecting anesthetic into the muscle, the syringe should be

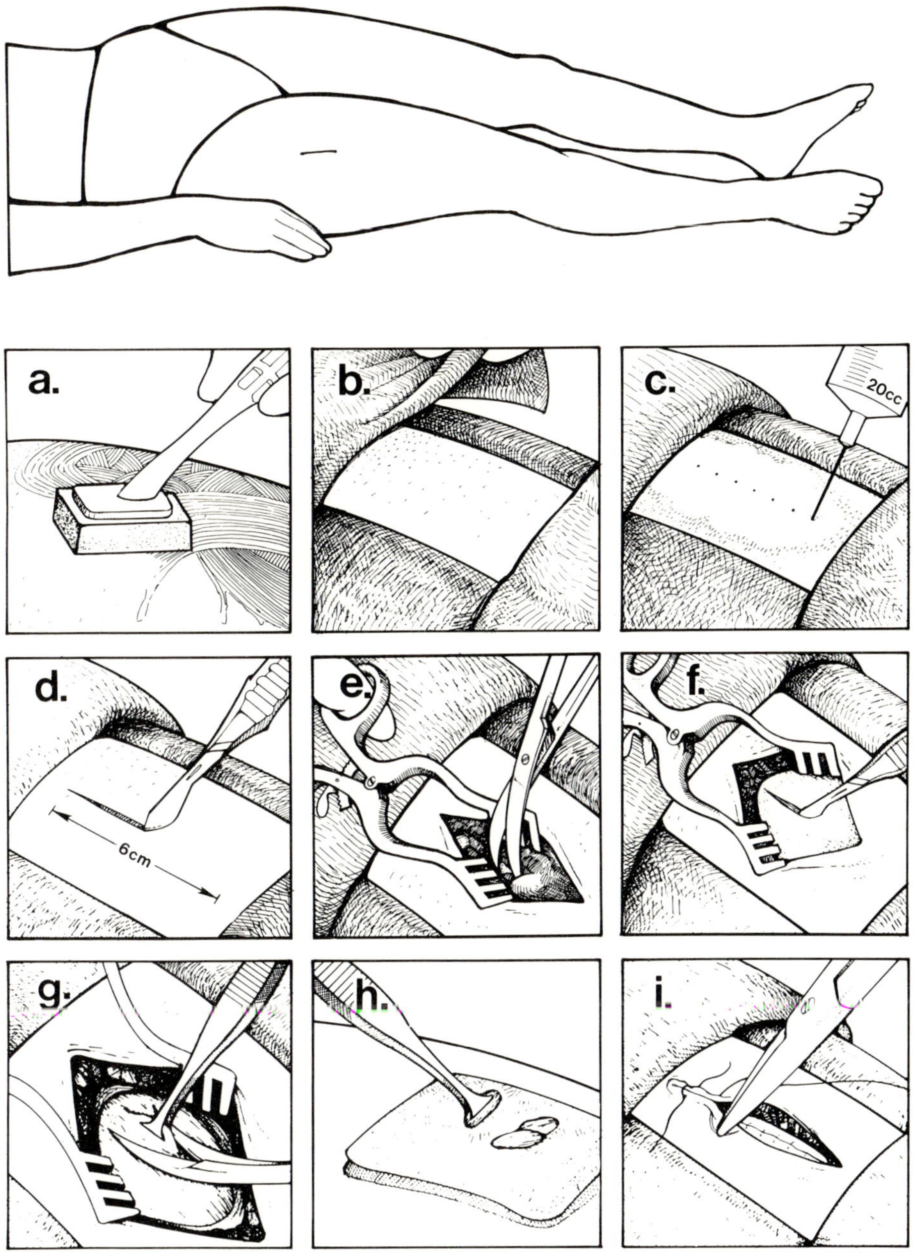

Figure 1 Technique of open muscle biopsy.

held almost parallel to the skin surface, and after insertion the needle should be used to slightly raise the skin as the anesthetic is injected. Usually, multiple punctures over a 6- to 10-cm length are necessary to instill the full amount of anesthetic and to deaden sufficient skin area for the incision.

The incision should be made parallel to the orientation of muscle fibers in the muscle belly. Sometimes, in cases of suspected inflammatory muscle disease, it is preferable to obtain a skin biopsy at the same time as the muscle biopsy. (A sliver of skin at the incision site should not be taken for the skin biopsy, because this will result in a ragged closure and greater scarring. Instead, a separate site for skin biopsy *away* from the skin incision for muscle biopsy should be used.) After the skin is incised with a scalpel, the subcutaneous tissue should be bluntly dissected with scissors. Cutting should be avoided as much as possible in order not to damage small superficial vessels and nerves in the area. A self-retaining retractor (for example, the Weitlaner retractor #SU3110, or the Beckman retractor #SU3195, American V. Mueller) should be used to separate the skin edges and maintain the exposure. Ligation or cautery of small bleeding vessels is rarely necessary, as these self-retaining retractors will apply enough pressure to the skin edges to stop the bleeding. Blunt dissection should continue down to the muscle fascia enveloping the muscle belly. The muscle fascia should be carefully incised and the muscle belly exposed. Once the muscle belly is exposed, it is necessary to avoid excessive trauma to the muscle fibers. Nontraumatic forceps (such as the O'Brien fixation forceps #OP3040, American V. Mueller) oriented parallel to the muscle fibers should be used to pick up the muscle and then, with scissors, a specimen can be cut parallel to the orientation of the muscle fibers. This should be done quickly, as the patient may experience a sharp twinge of pain. Any spurting of blood should be controlled by careful pressure with a gauze sponge until excessive bleeding stops.

Several specimens from different areas of the exposed muscle belly can be removed for study. The pieces of muscle should be placed carefully on a dry gauze sponge, which should then be placed on top of another sponge which has been moistened with several drops of normal saline (*not* saturated with fluid) in a dry petri dish. Allowing the muscle specimen to bathe in excess fluid will cause artifactual swelling of the endomysial connective tissue which results in separation and distortion of muscle fibers. There is no need to stretch the muscle sample. The cover should be placed on the petri dish and the samples should not be allowed to dry out or be exposed to heat. Heat damage of all types, such as from hot instruments, must be avoided, since it will destroy enzyme activity and prevent histochemical staining. Preferably, the samples should be taken to the laboratory immediately for processing, but if necessary a delay of several hours is possible if the samples are keep cool (about 8° to 14°C).

In some instances it may be of value to obtain samples for electron microscopy and/or biochemical analysis in addition to the muscle sample(s) for histochemical study. The sample for electron microscopy can be obtained by using a special clamp (the Vick–Olson clamp).* The clamp is applied to the muscle parallel to the orientation of the muscle fibers, and the strip of muscle attached to the prongs of the clamp is cut out. The clamp and attached muscle strip are then immediately plunged into a jar (a slide staining jar is a convenient size) of cold glutaraldehyde fixative† (Figure 2). The sample is allowed to fix for 3 to 4 hours while attached to the clamp. The sample is then diced into small pieces, returned to the fixative, and sent to an electron microscopy laboratory for further processing. Samples for biochemical study should be removed in the same manner as the pieces for histochemistry, but should be immediately dropped onto dry ice in a polystyrene bucket. These frozen specimens should then be placed into airtight containers in an ultracold (−80°C) freezer until the biochemical analysis is undertaken.

Once the muscle biopsy samples are obtained, the muscle fascia edges must be carefully opposed with suture. This is done to prevent the development of painful herniation of the muscle through a fault in the fascial layer. Synthetic absorbable suture material (such as Dexon sutures) is preferred to cat gut sutures, since there is less chance for an adverse inflammatory reaction to the suture material. Subsequently, the subcutaneous tissue and finally the

---

*Details of the construction and use of this clamp can be found in Vick NA, Olson WH: A biopsy clamp for removal and fixation of muscle in electron microscopy. *Stain Technology* 1970;45:42.

†Since glutaraldehyde penetrates tissue very slowly, a fixative containing a mixture of paraformaldehyde and glutaraldehyde gives better overall fixation for electron microscopy than glutaraldehyde alone. Either modified McDowell's fixative or half-strength Karnovsky's fixative may be used. Modified McDowell's fixative (1% glutaraldehyde and 4% paraformaldehyde) is made by dissolving 40 g of paraformaldehyde powder, 11.6 g of monobasic sodium phosphate (NaH_2PO_4), and 3.2 g of sodium hydroxide pellets in 980 mL of distilled water, heating to 60° to 70°C, and stirring until the paraformaldehyde is completely dissolved. After the mixture has cooled to room temperature, add 20 mL of electron microscopy grade 50% glutaraldehyde and adjust the solution pH to 7.4 using 1 mol/L sodium hydroxide or 1 mol/L hydrochloric acid. Half-strength Karnovsky's fixative (2.5% glutaraldehyde and 2% paraformaldehyde) is made by dissolving 20 g of paraformaldehyde powder in 250 mL of distilled water, heating to 60° to 70°C, and stirring until the paraformaldehyde is completely dissolved. Then while stirring, add 10 to 30 drops of 1 mol/L sodium hydroxide until the solution clears (slight milkiness may persist). After the mixture has cooled to room temperature, add 50 mL of electron microscopy grade 50% glutaraldehyde, then bring the volume to 1000 mL with 0.2 mol/L phosphate-buffered saline (8.68 g of dibasic sodium phosphate, 1.85 g of monobasic potassium phosphate, and 4.45 g of sodium chloride in 1000 mL of distilled water, adjusted to pH 7.4 with 1 mol/L sodium hydroxide or 1 mol/L hydrochloric acid). Glutaraldehyde containing solutions must be stored in the refrigerator to prevent deterioration. With Karnovsky's fixative there is better preservation of fine ultrastructural details, while with the modified McDowell's fixative larger pieces of tissue are more adequately fixed and there is greater contrast apparent in the microscopic sections. A convenient methodology for processing the fixed tissue for electron microscopy is the "pop-off" technique using the di Sant'Agnese-Jensen stain on Spurr epoxy embedded large sections (*Am J Clin Pathol* 1984;80:25–29; *Ultrastruct Pathol* 1984;6:247–253).

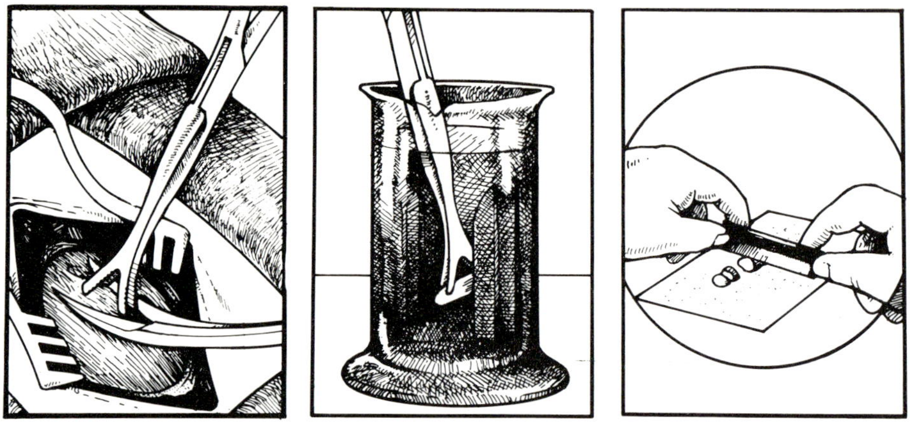

Figure 2 Obtaining specimens for electron microscopy during open muscle biopsy.

skin should be carefully sutured. Careful attention must be given to postbiopsy wound healing, especially in patients with dermatomyositis or patients receiving corticosteroid therapy. In such individuals, skin healing is compromised; skin sutures may need to remain longer and chronic open (nonhealing) surgical sites are possible.

NEEDLE BIOPSY

The needle biopsy technique is less elaborate, faster to perform, and results in less scarring than the open-biopsy technique. Samples from multiple muscles may be taken during one session. However, a significant drawback is the small sample size which may be inadequate for diagnosis. Generally, in infants and young children whose muscle fibers are of a much smaller diameter, the needle biopsy sample will contain more than enough muscle fibers to make an adequate diagnosis, but needle biopsy samples from adults will often contain less than 50 fibers and may make diagnosis unreliable.

To perform a needle muscle biopsy, the Tru-cut biopsy needle is recommended. A particularly apprehensive patient may need sedation with a mild anxiolytic-hypnotic agent. The biopsy area should be thoroughly cleansed and draped. About 5 mL of local anesthetic (lidocaine) should be injected subcutaneously. With a sterile technique, the biopsy needle should be opened and the trocar removed from the sheath. It is then necessary to reinsert the trocar into the sheath and manipulate the biopsy needle by alternately advancing the trocar and the sheath to be certain that there is smooth movement of the needle parts. The end of the trocar should then be bent slightly, as demonstrated in Figure 3. This maneuver will assure a more adequate muscle sample. With a scalpel point, a puncture the size of the tip of the biopsy needle should be made in the anesthetized skin. The biopsy needle is inserted perpendicular to the skin through this puncture hole and into the muscle belly. Once the tip is in the muscle belly, the handle of the biopsy needle is lowered to the skin surface so

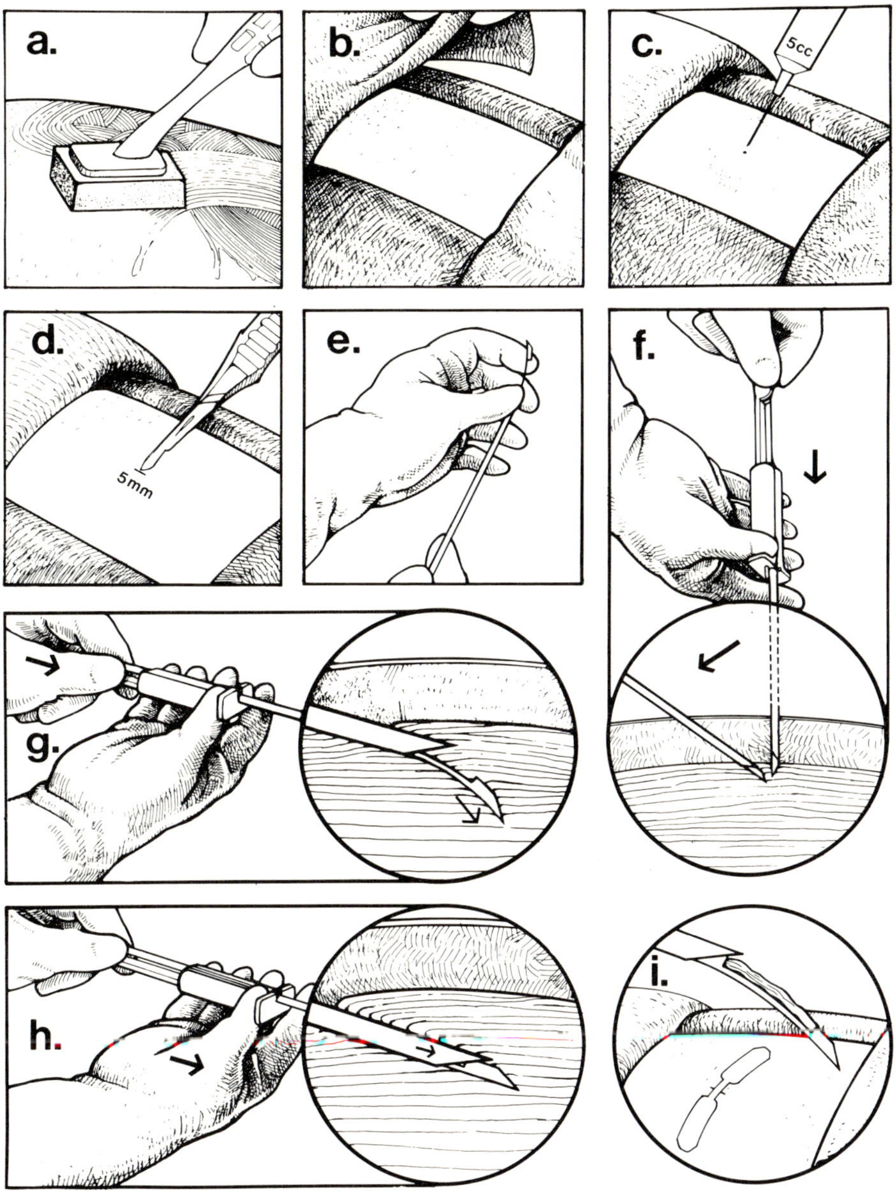

Figure 3 Technique of needle muscle biopsy.

that the barrel of the needle is essentially parallel with the orientation of muscle fibers in the muscle belly. The trocar tip is advanced; then the sheath is quickly advanced over the trocar tip and the needle removed. These latter maneuvers should be performed rapidly because the patient will experience some discomfort. As soon as the needle is removed from the skin, pressure should be applied over the skin opening and needle track to prevent excessive bleeding. The biopsy needle should be opened (by advancing the trocar) and the piece (or pieces) of muscle carefully removed onto a dry gauze sponge for subsequent processing.

Ideally, the needle biopsy specimen will consist of a single strip of muscle containing longitudinally oriented muscle fibers. Occasionally, multiple needle insertions into the same muscle belly can be made, but this is usually very painful for the patient and subsequent samples may be distorted by the previous biopsy trauma and bleeding. When the needle biopsy is completed, nonallergic tape should be applied across the puncture site to oppose the skin edges for healing with minimal scarring.

For several hours following open biopsy or needle biopsy, the patient should be encouraged to elevate the extremity, place an ice bag over the muscle, and avoid using the muscle. This will reduce the amount of bleeding (bruising) within the muscle and the amount of pain (soreness) during recovery.

Following the biopsy-induced trauma of the muscle, there is a general regeneration of muscle fibers. Patients should be encouraged to resume normal activity on the day following the biopsy, even if there is some muscle pain, in order to prevent the development of disuse atrophy.

2 HISTOCHEMICAL STAINING OF MUSCLE BIOPSY SPECIMENS

Once muscle biopsy specimens arrive at the laboratory, they must be quickly processed to prevent damage which may produce artifacts. Ideally, the muscle specimens should be quick-frozen within 30 minutes of removal from the patient. Specimens processed after 30 minutes show increasing loss of enzymatic staining. Prior to sectioning, the muscle biopsy must be quick-frozen onto a microtome chuck. The preferred cryostat is a Tissue-Tek II. For the cryostat microtome, readymade plastic chucks may be purchased (Arthur H. Thomas Co, Philadelphia, PA, #6755-B10). Alternatively, chucks can be made by sawing 5/8-inch thick sections from 1-inch wood doweling.

In processing the muscle biopsy, the specimen must first be properly trimmed and oriented (Figure 4). A hand magnifying lens is helpful in this procedure. As atraumatically as possible, the muscle specimen should be picked up at one end with forceps and transferred from the petri dish gauze to the surface of a dental wax sheet (Sealon Pink Base Plate Wax No. 2, Howmedica Inc, Dental Division). Using the hand magnifier, the orientation of the muscle fibers should be determined. A new, sharp razor blade (WEC PREP Prep Blades, WECK #450205) can be used to trim the specimen edges parallel to the fiber orientation to produce a cylindrical specimen of about 1 cm diameter and 2 to 3 cm length. Gum tragacanth is used to affix the specimen to the chuck. Gum tragacanth may be prepared according to the following formula:

Gum tragacanth	2 g
Distilled water	18 mL
Phenol	3 drops

The addition of phenol prevents microorganism growth. The gum tragacanth is mounded onto the chuck with a small spatula. A hole is made in the center of the mound for the muscle specimen. The muscle specimen should be placed on the chuck so that the muscle fibers are oriented upright (in this orientation the sections will be transverse sections of the muscle fibers). It may be desirable to mount two muscle pieces next to each other, one placed upright to yield transverse sections and the other placed horizontally to yield longitudinal sections. The gum tragacanth should be molded completely around the bottom one-third of the muscle specimen. The upper two-thirds of the muscle specimen must be above the level of the gum tragacanth; otherwise, when sectioning in the microtome, the knife blade will be damaged from cutting into the

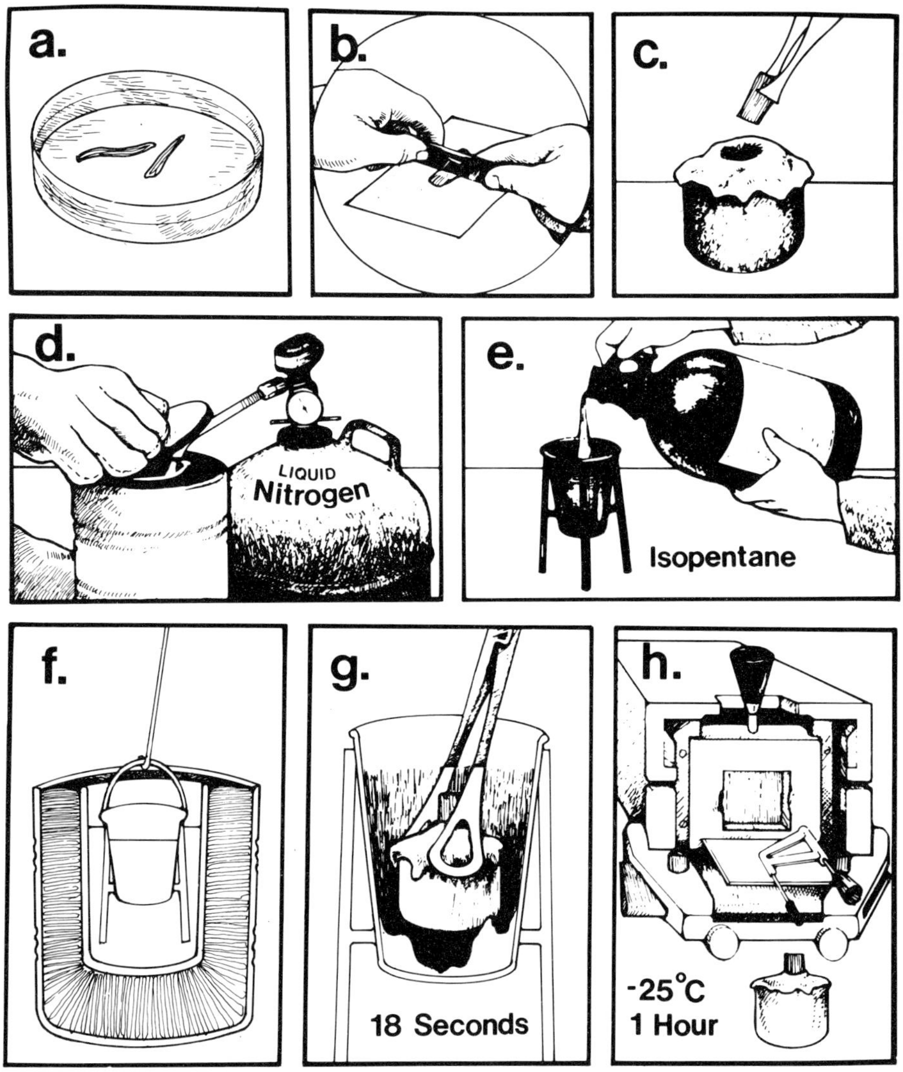

Figure 4 Technique of freezing muscle specimen.

frozen gum tragacanth. It is preferable to place no more than two muscle specimens on a single chuck (ideally, the cross-sectional measurements of specimen on a single chuck should not exceed 10 mm). Clean instruments should be used to handle the muscle specimen because water droplets may promote ice artifact.

The freezing is accomplished in isopentane (2-methylbutane) cooled to liquid nitrogen temperature. If the muscle were immersed directly into the −190°C liquid nitrogen, an insulating layer of nitrogen gas would form around the muscle, with resultant slow freezing. Instead, the muscle specimen is placed in isopentane, which is a heat-conducting medium and results in rapid freezing of the specimen. Rapid freezing is necessary because slower freezing produces ice artifact.

A Dewar flask (Figure 5) is filled half full with liquid nitrogen. A wide-mouth metal cup or beaker is filled three-fourths full with isopentane. After the liquid nitrogen has stopped bubbling in the Dewar flask, the beaker containing isopentane is immersed very carefully and slowly. Cover the flask to improve cooling. When the isopentane reaches −160°C, it becomes slightly syrupy and has whitish globs of solid isopentane at the side or on the bottom of the beaker.*

To freeze the muscle specimen, the chuck is held by long forceps with the muscle specimen vertical (the muscle should be straight). The chuck is plunged rapidly into the isopentane and held 10 to 20 seconds. Immersing the specimen for too short a period results in incomplete freezing with

*Some laboratories have found that alternate containers have proven satisfactory for the isopentane and the liquid nitrogen. A large glass Pyrex beaker has been used to hold the isopentane, and a large leakproof polystyrene container to hold the liquid nitrogen. A slightly less desirable technique for freezing the muscle specimens is to cool a container (glass beaker or metal cup) of isopentane in an acetone-dry ice bath (or solid dry ice) and immerse the biopsy in the cooled isopentane (still in the dry ice).

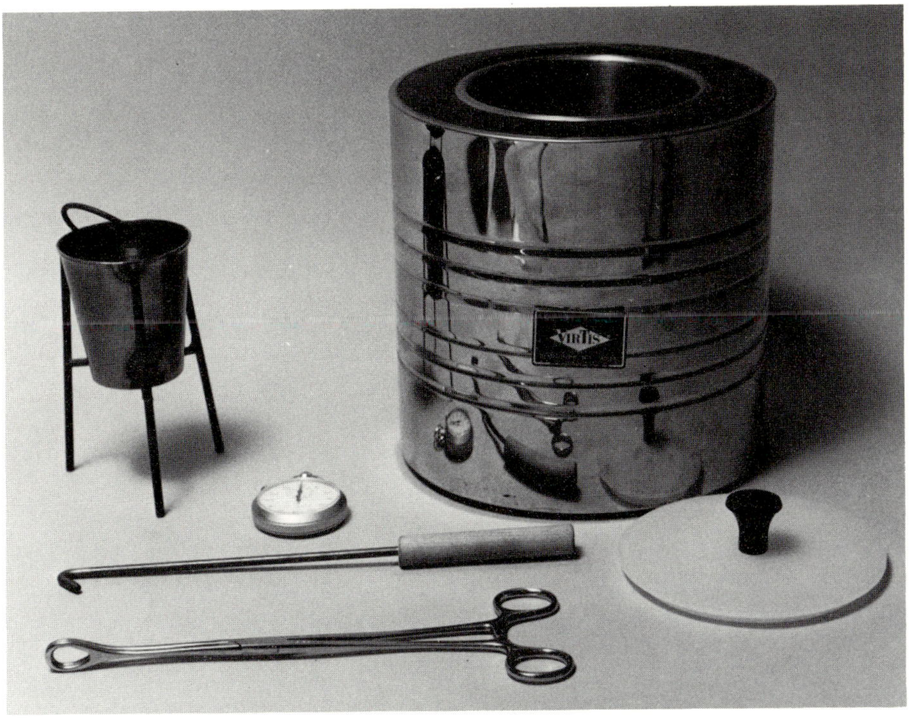

Figure 5 Equipment necessary to freeze muscle specimens.

ice artifact evident in the center of the specimen; immersion for too long a period will cause the block to crack off the chuck.

Once the specimen is frozen, it should be removed from the isopentane. Shake the specimen once or twice to remove excess isopentane, then place it immediately into the −20° to −25°C cryostat or into an ultracold (−80°C) freezer. The specimen must not be allowed to thaw because ice artifact will develop. The specimen should not be placed in a closed container immediately after freezing because the isopentane must be allowed to evaporate from the specimen. Evaporation usually takes 1 hour at cryostat temperature. In addition, if the block is sectioned before the isopentane has evaporated, there will be an excessive number of wrinkles and bubbles. Specimens can be stored long-term in airtight plastic containers placed in an ultracold freezer. Such containers must be precooled before specimens are placed in them.

Sections should be cut only when the chuck and specimen have equilibrated to the cryostat temperature, which should be from −20° to −25°C. Sections can be cut at 10 μm thickness and picked up onto square #2 thickness coverslips. The sections are then air dried on the coverslips. For most stains, the coverslips may be left overnight before staining; however, the SDH stain must be performed within 5 minutes, and the modified Gomori trichrome stain must be done within 1 hour of sectioning.

To perform histochemical stains, the coverslips should be placed in a small coverslip jar or a coverslip rack which can be immersed in Columbia staining jars. When staining is complete, the coverslips are mounted onto slides. A synthetic mounting medium (such as Permount) or glycerine jelly is usually used. The glycerine jelly is made as follows:

Combine

Gelatin	10 g
Distilled water	60 mL

Heat until gelatin is dissolved, then add

Glycerine	70 mL
Phenol	1 mL

The glycerine jelly solution will remain usable up to 6 weeks in the refrigerator.

Because many of these stains (especially the ATPase) fade with time, representative changes in the muscle may be photographed for a permanent record.

Directions for various staining methods are shown below:

Hematoxylin and Eosin (H&E)

SOLUTIONS

Harris' Hematoxylin

hematoxylin crystals	5 mg
absolute alcohol	50 mL
potassium aluminum sulfate (alum)	100 g
distilled water	1000 mL
mercuric oxide (red)	2.5 g

Dissolve hematoxylin in alcohol. Dissolve potassium alum in hot water. Remove alum solution from heat and mix the two solutions. Then bring to boil as rapidly as possible. Remove from heat and add mercuric oxide slowly. Reheat until solution becomes dark purple, remove immediately, and plunge the container into ice water until cold. The solution is ready for use as soon as it is cool. Filter before use. (Harris' hematoxylin may be purchased ready-made through S/P Harleco #683.)

Lithium Water

distilled water	100 mL
lithium carbonate	1 g

Alcoholic Eosin

eosin Y	6 g
70% ethanol	600 mL
orange G	1.5 g
acid fuchsin	0.75 g
glacial acetic acid	0.33 mL

Dissolve stains in the alcohol. Add acetic acid last.

STAINING PROCEDURE

1. Place sections in Harris' hematoxylin solution for 5 minutes.
2. Wash in tap water to remove excess hematoxylin.
3. Place sections in lithium water for 1 minute.
4. Wash in tap water for 2 minutes.
5. Dip two times in 80% alcohol.
6. Stain in alcoholic eosin solution for 3 minutes.
7. Dehydrate in graded alcohols and clear in xylene (95% alcohol, 95% alcohol, 100% alcohol, 100% alcohol-xylene, xylene, xylene). Mount in synthetic medium.

Modified Gomori Trichrome

SOLUTIONS

Trichrome Solution

fast green FCF	0.3 g
chromotrope 2R	0.6 g
phosphotungstic acid	0.6 g
glacial acetic acid	1 mL
distilled water	100 mL

Mix well. Adjust pH to 3.4.

Harris' Hematoxylin (See H&E Stain)

STAINING PROCEDURE

1. Place sections in Harris' hematoxylin for 5 minutes.
2. Wash in distilled water to remove excess hematoxylin.
3. Stain in trichrome solution for 20 minutes.
4. Wash in distilled water to remove excess trichrome.
5. Dip twice in 0.2% acetic acid.
6. Dehydrate in graded alcohols and clear in xylene. Mount in synthetic medium.

Comment: Staining must be done within 1 hour of cutting sections.

Alkaline Myofibrillar Adenosine Triphosphatase (ATPase)*

SOLUTIONS

Sodium Barbital Buffer

sodium barbital	4.124 g
calcium chloride	1.998 g
distilled water	1000 mL

Adjust pH to 9.4.

---

*In some laboratories an alternate method is used (*Histochem J* 1980;12:707–710). Instead of the sodium barbital buffer, a glycine buffer is used consisting of 3.75 g glycine, 2.9 g sodium chloride, and 11.1 g calcium chloride dissolved in 600 mL distilled water and mixed with 350 mL of 0.1 mol/L sodium hydroxide; the pH is then adjusted to 9.5 with 1 mol/L sodium hydroxide. The incubation solution consists of a mixture of 5 mg adenosine-5-triphosphate (ATP), 10 mL glycine buffer, and 1 drop of 1 mmol/L dithiothreitol (1.5 mg in 10 mL distilled water). Because dithiothreitol is harmful to pH electrodes, only the pH of the glycine buffer is checked immediately prior to mixing the incubation solution. There is also an additional six minute washing with a 1% calcium chloride solution (10 g calcium chloride in 1000 mL distilled water) immediately after the incubation and prior to transfer to the 2% cobaltous chloride solution.

2% Cobaltous Chloride

cobaltous chloride ($CoCl_2$)	20 g
distilled water	1000 mL

Stoer-Ammonium Sulfide (Light Solution)
$(NH_4)_2S$ + AQ

ammonium sulfide	0.4 mL
tap water	10 mL

Incubation Solution

sodium barbital buffer	10 mL
adenosine-5-triphosphate (ATP)	15 mg

Adjust pH of incubation solution to 9.4.

STAINING PROCEDURE

1. Place sections in sodium barbital buffer for 10 minutes.
2. Incubate in incubation solution at 37°C for 30 minutes (15 to 20 minutes for non-human animal muscle).
3. Transfer to 2% cobaltous chloride solution for 3 minutes.
4. Wash in distilled water. Change distilled water every 30 seconds for 5 minutes.
5. Transfer to ammonium sulfide solution for 10 seconds.
6. Wash in running tap water for 20 minutes.
7. Dehydrate in graded alcohols and clear in xylene. Mount in synthetic medium.

Comment: The Stoer-ammonium sulfide solution and incubation solution should be made fresh each time.

Myofibrillar Adenosine Triphosphatase— EDTA Reversal

SOLUTIONS

EDTA Buffer

EDTA (disodium salt)	3.72 g
distilled water	1000 mL

Adjust pH to 4.35.

Sodium Barbital Buffer (See ATPase Stain)

2% Cobaltous Chloride (See ATPase Stain)

Stoer-Ammonium Sulfide (Light Solution) (See ATPase Stain)

STAINING PROCEDURE

1. Preincubate sections in EDTA buffer at room temperature for 1 hour.
2. Transfer to sodium barbital buffer for 3 minutes.
3. Stain as for alkaline ATPase reaction, joining the alkaline ATPase procedure at step 2.

Nicotinamide Adenine Dinucleotide (Reduced)-Tetrazolium Reductase (NADH-TR)

SOLUTIONS

Gomori TRIS-HCl Buffer, pH 7.4

tris(hydroxymethyl)aminomethane (THAM) base	6.06 g
concentrated HCl	3.34 mL
distilled water	1000 mL

Incubation Solution

Gomori TRIS-HCl buffer	10 mL
nitro blue tetrazolium (NBT)	10 mg
β-nicotinamide adenine dinucleotide, reduced form (NADH)	4 mg

Adjust pH to 7.4.

STAINING PROCEDURE

1. Incubate sections in incubation solution at 37°C for 30 minutes.
2. Wash in distilled water for 1 minute.
3. Dehydrate in 30%, 60%, 90% acetone, then rehydrate in 60%, 30% acetone.
4. Rinse in distilled water for 1 minute.
5. Mount in glycerine jelly.

Comment: The NADH (β-nicotinamide adenine dinucleotide) is stored at room temperature in the dark with dessicator. This stain was formerly called DPNH-diaphorase.

Succinic Dehydrogenase (SDH)

SOLUTIONS

0.2 mol/L Sodium Succinate

sodium succinate	32.4 g
distilled water	1000 mL

0.2 mol/L Phosphate Buffer (pH 7.4)

sodium phosphate, dibasic, anhydrous	22.72 g
potassium phosphate, monobasic	5.40 g
distilled water	1000 mL

Incubation Solution

0.2 mol/L sodium succinate	5 mL
0.2 mol/L phosphate buffer	5 mL
nitro blue tetrazolium (NBT)	10 mg

Adjust pH to 7.2–7.6.

STAINING PROCEDURE

1. Incubate sections in incubation solution at 37°C for 1 hour.
2. Wash in distilled water for 1 minute.
3. Dehydrate in 30%, 60%, 90% acetone, then rehydrate in 60%, 30% acetone.
4. Rinse in distilled water for 1 minute.
5. Mount in glycerine jelly.

Comment: The sections should be air dried for 5 minutes after cutting, then transferred immediately into the SDH incubation solution.

Nonspecific Esterase

SOLUTIONS

0.2 mol/L Phosphate Buffer, pH 7.4 (See SDH Stain)

Pararosanilin Stock Solution

pararosanilin hydrochloride	1 g
distilled water	20 mL
concentrated HCl	5 mL

Warm solution slowly. Cool to room temperature. Filter and store in refrigerator. It will remain stable for weeks or until depleted.

4% Sodium Nitrite

sodium nitrite ($NaNO_2$)	4 g
distilled water	100 mL

1% Alpha-Naphthyl Acetate in Acetone

alpha-naphthyl acetate	1 g
acetone	100 mL

Azotized Pararosanilin

4% sodium nitrite	0.8 mL
pararosanilin stock solution	0.8 mL

Add pararosanilin *to* sodium nitrite. (Will precipitate if mixed in other direction.)

Incubation Solution

Mix in this order

0.2 mol/L phosphate buffer, pH 7.4	20 mL
1% alpha-naphthyl acetate	0.5 mL
azotized pararosanilin	1.6 mL

Adjust pH to 6.8–7.2.

STAINING PROCEDURE

1. Incubate sections in incubation solution for 1 hour at 37°C.
2. Wash sections for 10 minutes in tap water.
3. Dehydrate sections in graded alcohols and clear in xylene. Mount in synthetic medium.

Comment: The slides may become very sticky with a granular deposit. Before mounting, wipe off with a tissue dipped in xylene. The sodium nitrite is very unstable and should be made just before use.

Alkaline Phosphatase

SOLUTIONS

0.1 mol/L Sodium Barbital

sodium barbital	2.06 g
distilled water	100 mL

Incubation Solution

sodium α-naphthyl acid phosphate	25 mg
fast blue RR salt	25 mg
0.1 mol/L sodium barbital	25 mL

Adjust pH to 9.2.

STAINING PROCEDURE

1. Incubate sections in incubation solution at room temperature for 1 hour.
2. Wash 3 minutes in distilled water.
3. Wash 2 minutes in 1% acetic acid.
4. Rinse in distilled water.
5. Temporarily mount in glycerine for 1 week while reaction bubbles continue to develop.
6. Remount in glycerine jelly.

Menadione-Linked Alpha-Glycerophosphate Dehydrogenase

SOLUTIONS

Gomori TRIS-HCl Buffer, pH 7.4

tris(hydroxymethyl)aminomethane (THAM) base	6.06 g
concentrated HCl	3.34 mL
distilled water	1000 mL

Incubation Solution

Gomori TRIS-HCl buffer	10 mL
menadione	4 mg
nitro blue tetrazolium (NBT)	10 mg
alpha-glycerophosphate	30 mg

Adjust pH to 7.4.

STAINING PROCEDURE

1. Incubate sections in incubation solution at 37°C for 1 hour.
2. Wash in distilled water for 1 minute.
3. Dehydrate in 30%, 60%, 90% acetone, then rehydrate in 60%, 30% acetone.
4. Rinse in distilled water for 1 minute.
5. Mount in glycerine jelly.

Periodic Acid–Schiff (PAS)

SOLUTIONS

0.5% Periodic Acid

periodic acid	0.5 g
distilled water	100 mL

Schiff's Solution

basic fuchsin	1 g
distilled water	200 mL
1 mol/L HCl	20 mL
sodium metabisulfite	1 g

Dissolve basic fuchsin in hot, distilled water. Bring to boilin point. Cool to 50 °C. Filter and add 1 mol/L HCl. Cool further and add sodium metabisulfite. Keep in dark for 48 hours until solution becomes straw colored. Store in refrigerator. (Schiff's solution may be purchased readymade through S/P Harleco #6073.)

Note: Keep Schiff's solution in refrigerator, but bring to room temperature before using.

Harris' Hematoxylin (See H&E Stain)

STAINING PROCEDURE

1. Transfer sections to 0.5% periodic acid solution for 2 to 5 minutes.
2. Wash in distilled water for 1 minute.
3. Transfer to Schiff's solution for 15 minutes.
4. Wash in running tap water for 10 minutes.
5. Counterstain in Harris' hematoxylin for 1 to 2 minutes.
6. Wash in tap water for 5 minutes.
7. Dehydrate in graded alcohols and clear in xylene. Mount in synthetic medium.

Comment: A negative control may be run by preincubating slides in malt diastase.

SOLUTION

malt diastase	0.5 g
distilled water	100 mL

PROCEDURE

1. Wash sections in distilled water.
2. Transfer sections to malt diastase solution for 1 hour at 37°C.
3. Wash sections well in tap water for 5 to 10 minutes.
4. Transfer sections to 0.5% periodic acid solution and stain as described above.

Alizarin Red S

SOLUTIONS

1% Alizarin Red S

alizarin red S	667 mg
distilled water	60 mL

Mix thoroughly. Add 6.7 mL of NH_4 solution (28% ammonium water 0.1 mL, plus distilled water 9.9 mL) while stirring. Adjust pH to 6.3–6.5.

STAINING PROCEDURE

1. Stain section in 1% alizarin red S for 2 minutes. Timing can be critical, especially in nonhuman animal muscle. *Do not overstain.*
2. Drain and flush sections three to four times using acetone for 1 minute.
3. Transfer to acetone:xylene (1:1) for 1 minute, then clear in two changes of xylene.
4. Mount directly in synthetic medium.

Adenylate Deaminase

SOLUTIONS

3 mol/L Potassium Chloride

potassium chloride	22.4 g
distilled water	100 mL

150 mmol/L KCl–1.5 mmol/L Citrate

potassium chloride	1.12 g
sodium citrate	32 mg
distilled water	100 mL

Adjust pH to 6.0.

Dithiothreitol Solution

dithiothreitol	5.0 mg
distilled water	0.3 mL

Incubation Solution

nitro blue tetrazolium (NBT)	10 mg
distilled water	9 mL
5′-adenylylimidodiphosphate (AMP-PNP)	4 mg
3 mol/L potassium chloride	0.7 mL

Adjust pH to 6.1.

STAINING PROCEDURE

1. After adjusting pH of incubation solution, add dithiothreitol solution with a dropper while stirring *just prior to staining.*
2. Incubate sections in incubation solution for 1 hour at room temperature.
3. Rinse coverslips twice with 150 mmol/L KCl–1.5 mmol/L citrate solution and mount in glycerine jelly.

Comment: The pH of the incubation solution is adjusted *before* addition of the dithiothreitol, because this compound may damage pH electrodes.

Toluidine Blue

SOLUTIONS

0.5% Toluidine Blue

toluidine blue	0.5 g
20% ethanol	100 mL

Combine. When dye is completely dissolved, filter. Stain improves when aged.

STAINING PROCEDURE

1. Place sections in toluidine blue for 20 minutes at room temperature.
2. Wash briefly in tap water.
3. Dehydrate in graded alcohols and clear in xylene. Mount in synthetic medium.

Comment: This is a metachromatic staining procedure.

Oil Red O

SOLUTIONS

Stock Saturated Solution

oil red O	0.5 g
isopropanol (isopropyl alcohol)	100 mL

This stock saturated solution is stable.

Working Solution (combine just before use)

oil red O stock solution	12 mL
distilled water	8 mL

Harris' Hematoxylin (See H&E Stain)

STAINING PROCEDURE

1. Prepare the working solution of oil red O. Let stand for 5 to 10 minutes, then filter. The filtrate is usable for several hours.
2. Place sections in working solution for 10 to 15 minutes at room temperature.
3. Wash in distilled water.
4. Stain nuclei briefly (10 to 30 seconds) in Harris' hematoxylin.
5. Wash in distilled water.
6. Mount in glycerine jelly. (*Caution:* The glycerine jelly should only be slightly warm or it will melt the fat in the section causing artifact.)

Sudan Black B

SOLUTIONS

Sudan Black B

Sudan black B	0.7 g
propylene glycol, 100%	100 mL

Dissolve Sudan black B in propylene glycol by heating to 100°–110°C and stirring thoroughly for a few minutes. *Do not exceed 110°C.* Filter hot through Whatman No. 2 paper to remove excess dye. After cooling to room temperature, filter again.

85% Propylene Glycol

propylene glycol, 100%	85 mL
distilled water	15 mL

Harris' Hematoxylin (See H&E Stain)

STAINING PROCEDURE

1. Place sections in 100% propylene glycol for 10 minutes at room temperature.
2. Incubate in Sudan black B overnight at room temperature.
3. Differentiate in 85% propylene glycol for 3 minutes. Agitate several times.
4. Rinse in distilled water.
5. Counterstain in Harris' hematoxylin for 5 minutes.
6. Rinse well in distilled water.
7. Mount with glycerine jelly. (*Caution:* The glycerine jelly should only be slightly warm or it will melt the fat in the section causing artifact.)

Acid Phosphatase

SOLUTIONS

Barbiturate Acetate Buffer

sodium acetate	9.7 g
sodium barbiturate	14.7 g
distilled water	500 mL

Azotized Pararosanilin (see Nonspecific Esterase Stain)

0.2 mol/L Sodium Phosphate Solution

disodium biphosphate, anhydrous (Na_2HPO_4)	20.7 g
distilled water	1000 mL

Acetone Fixative

acetone	60 mL
barbiturate acetate buffer	40 mL

Incubation Solution

Mix in this order:

naphthol AS-BI phosphate (sodium salt)	10 mg
N,N-dimethylformamide	1 mL
barbiturate acetate buffer	5 mL
distilled water	12 mL
azotized pararosanilin	1.6 mL

Adjust pH to 5.0.

Methyl Green Counterstain

methyl green	1 g
0.2 mol/L sodium phosphate solution	100 mL

STAINING PROCEDURE

1. Fix sections in acetone fixative for 3 to 4 minutes at 4°C.
2. Wash in distilled water.
3. Incubate sections in incubation solution at 37°C for 1 hour.
4. Wash briefly in distilled water.
5. Counterstain in methyl green for several seconds.
6. Temporarily mount in glycerine for 1 week.
7. Remount in glycerine jelly.

Lactate Dehydrogenase (LDH)

SOLUTIONS

0.2 mol/L Phosphate Buffer (See SDH Stain)

85% Lactic Acid

Incubation Solution

0.2 mol/L phosphate buffer	10 mL
85% lactic acid	0.1 mL
β-nicotinamide adenine dinucleotide, oxidized form (NAD)	4 mg
nitro blue tetrazolium (NBT)	10 mg

Adjust pH of incubation solution to 7.4.

STAINING PROCEDURE

1. Incubate sections in incubation solution at 37°C for 1 hour.
2. Wash in distilled water for 1 minute.
3. Dehydrate in 30%, 60%, 90% acetone, then rehydrate in 60%, 30% acetone.
4. Rinse in distilled water.
5. Mount in glycerine jelly.

Nicotinamide Adenine Dinucleotide Phosphate (Reduced)-Tetrazolium Reductase (NADPH-TR)

SOLUTIONS

Gomori TRIS-HCl Buffer, pH 7.4

tris(hydroxymethyl)aminomethane (THAM) base	6.05 g
concentrated HCl	3.34 mL
distilled water	1000 mL

Incubation Solution

Gomori TRIS-HCl buffer	10 mL
nitro blue tetrazolium (NBT)	10 mg
β-nicotinamide adenine dinucleotide phosphate, reduced form (NADPH)	4 mg

Adjust pH of incubation solution to 7.4.

STAINING PROCEDURE

1. Incubate sections in incubation solution at 37°C for 20 to 30 minutes.
2. Wash in distilled water for 1 minute.
3. Dehydrate in 30%, 60%, 90% acetone, then rehydrate in 60%, 30% acetone.
4. Rinse in distilled water for 1 minute.
5. Mount in glycerine jelly.

Comment: This stain was formerly called TPNH-diaphorase.

Glyceraldehyde-3-Phosphate Dehydrogenase (GR-3-PD)

SOLUTIONS

0.2 mol/L Phosphate Buffer (pH 7.2)

sodium phosphate, dibasic, anhydrous	22.72 g
potassium phosphate, monobasic	5.40 g
distilled water	1000 mL

Incubation Solution (Stable for Several Months at −20°C)

glyceraldehyde-3-phosphate free acid (Sigma Chemical Co.)	0.25 mL
distilled water	6.35 mL
β-nicotinamide adenine dinucleotide, oxidized form (NAD)	7.5 mg
nitro blue tetrazolium (NBT)	6.0 mg
EDTA (disodium salt)	3.6 mg
0.2 mol/L phosphate buffer	3.0 mL

Adjust to pH of incubation solution to 7.2.

STAINING PROCEDURE

1. Incubate sections in incubation solution at 37°C for 20 minutes.
2. Rinse in distilled water for 1 minute.
3. Dehydrate in 30%, 60%, 90% acetone, then rehydrate in 60%, 30% acetone.
4. Rinse in distilled water for 1 minute.
5. Mount in glycerine jelly.

Amylophosphorylase

SOLUTIONS

0.2 mol/L Sodium Acetate Buffer

sodium acetate	27.2 g

1 mol/L HCl	40 mL
distilled water	960 mL

Lugol's Solution

iodine	1.0 g
potassium iodide	2.0 g
distilled water	100 mL

Dilute Lugol's Solution

Lugol's solution	1 mL
distilled water	9 mL

Iodine-Glycerine

Lugol's solution	2 mL
glycerine	20 mL

Incubation Solution

0.2 mol/L sodium acetate buffer	20 mL
glucose-1-phosphate (potassium salt)	25 mg
adenosine monophosphate (AMP)	5 mg
glycogen (shellfish)	2 mg
95% ethanol	2 mL

Adjust pH of incubation medium to 5.6.

STAINING PROCEDURE

1. Incubate sections in incubation solution at 37° C for 1 hour.
2. Remove sections and place directly into dilute Lugol's solution for 10 to 20 seconds.
3. Remove, blot excess iodine, and mount in iodine-glycerine.

Comment: This is not a permanent stain and will fade as the iodine sublimes. To restore stain, remove coverslip and place in diluted Lugol's solution until color is restored. Mount as before.

Adenyl Cyclase

SOLUTIONS

TRIS-HCl Buffer

tris(hydroxymethyl)aminomethane (THAM) base	1.211 g
distilled water	100 mL

Incubation Solution

TRIS-HCl buffer	10 mL
5′-adenylylimidodiphosphate (AMP-PNP)	2.65 mg
sodium fluoride (NaF)	2.1 mg
theophylline	3.6 mg
magnesium sulfate ($MgSO_4$)	10.0 mg
strontium chloride ($SrCl_2$)	53.4 mg

Adjust pH to 8.5.

Stoer–Ammonium Sulfide (Light Solution) (See ATPase Stain)

2% Cobaltous Chloride (See ATPase Stain)

STAINING PROCEDURE

1. Incubate sections in incubation solution for 48 hours at 25°C.
2. Wash in 2% cobaltous chloride for 3 minutes.
3. Rinse in distilled water (change every 30 seconds) for 5 minutes.
4. Transfer to ammonium sulfide solution for 30 seconds.
5. Rinse in running tap water for 15 minutes.
6. Dehydrate in graded alcohol and clear in xylene; mount in a synthetic mounting medium.

Guanidyl Cyclase

SOLUTIONS

TRIS-HCl Buffer

tris(hydroxymethyl)aminomethane (THAM) base	1.211 g
distilled water	100 mL

Incubation Solution

TRIS-HCl buffer	10 mL
5′-guanylylimidodiphosphate (GMP-PNP)	2.65 mg
sodium fluoride (NaF)	2.1 mg
theophylline	3.6 mg
magnesium sulfate ($MgSO_4$)	10.0 mg
strontium chloride ($SrCl_2$)	53.4 mg

Adjust pH to 8.5.

2% Cobaltous Chloride (See ATPase Stain)

Stoer–Ammonium Sulfide (Light Solution) (See ATPase Stain)

STAINING PROCEDURE

1. Incubate sections in incubation solution for 48 hours at 25°C.
2. Wash in 2% cobaltous chloride for 3 minutes.
3. Rinse in distilled water (change every 30 seconds) for 5 minutes.
4. Transfer to ammonium sulfide solution for 30 seconds.
5. Rinse in running tap water for 15 minutes.
6. Dehydrate in graded alcohols and clear in xylene. Mount in a synthetic mounting medium.

Wright's Stain

SOLUTIONS

Wright's Stain

Wright's stain	1.5 g
Giemsa's stain	0.15 g
glycerine	15 mL
methyl alcohol	500 mL

Grind the powdered stains and glycerine in a dry mortar. Add small amounts of methyl alcohol. Transfer to a clean brown bottle and let stand for 1 to 2 weeks, mixing daily.

Wright's Buffer

potassium phosphate, monobasic	6.63 g
sodium phosphate, dibasic	4.5 g
distilled water	1000 mL

Adjust pH to 6.4.

Methyl Alcohol

STAINING TECHNIQUE

1. Fix specimen with methyl alcohol for 2 minutes.
2. Flood specimen with filtered Wright's stain for 5 minutes.
3. Add an equal volume of Wright's buffer. To mix, blow over the surface of the specimen until a metallic green color appears. Let stand for 5 minutes.
4. Rinse with distilled water.
5. Air dry.
6. Mount in synthetic medium.

CHARACTERISTICS AND USE OF STAINS

These various stains are not all used for every muscle biopsy specimen. The standard histochemical battery for the initial evaluation of a muscle biopsy specimen consists of five stains – modified Gomori trichrome, H&E, alkaline myofibrillar ATPase, NADH-TR, and nonspecific esterase. Subsequently, other stains may be performed based upon the individual case. The staining characteristics and usefulness of the 23 stains discussed in this text are shown in Table 1 and described below.

Hematoxylin and Eosin (H&E)

This stains the nuclei of muscle fibers and other cells blue. The intermyofibrillar network of muscle cells also stains blue. The remainder of the muscle cell cytoplasm, other cells, and surrounding connective tissue stain pink. This stain is useful in assessing the overall basic features of the biopsy specimen, such as size and number of fibers, connective tissue, presence of inflammatory cells, muscle spindle, blood vessels, and nerve.

Modified Gomori Trichrome

This is the single most universally useful stain for morphological assessment of the muscle biopsy. Various components of the muscle fiber stain differentially with the nuclei purple, intermyofibrillar network red, and myofibrils green (except the Z-disc, which stains red). It is also frequently possible to differentiate muscle fiber types since the ATPase type I fibers stain darker than the ATPase type II fibers. Connective tissue stains light green; the spindle capsule and blood vessels stain darker green. The myelin sheath of nerves stains red; nerve axons stain purple.

Alkaline Myofibrillar ATPase

This is the standard stain for differentiating muscle fiber types. Only muscle fibers are stained, with type II muscle fibers staining dark brown and type I muscle fibers staining very pale to light brown. The smooth muscle of blood

Table 1
Muscle Fiber Type Differentiation

Stain	Type I	Type II
Modified Gomori Trichrome	Dark	Light
Alkaline Myofibrillar ATPase	Light	Dark
EDTA-Reversed Myofibrillar ATPase	Dark	Light
NADH-Tetrazolium Reductase	Dark	Light
Succinic Dehydrogenase	Dark	Light
Nonspecific Esterase	Dark	Light
Menadione-linked α-Glycerophosphate Dehydrogenase	Light	Dark
Periodic Acid–Schiff	Light	Dark
Oil Red O	Dark	Light
NADPH-Tetrazolium Reductase	Dark	Light
Amylophosphorylase	Light	Dark

vessel walls also stains darkly. The size and shape of individual fibers and the distribution of fiber types is determined with this stain.

EDTA-Reversed Myofibrillar ATPase

This stain is the reverse of the alkaline myofibrillar ATPase stain; that is, with EDTA preincubation the type I fibers stain dark brown and the type II fibers are nearly colorless. Usually, this stain is used only to confirm the fiber typing noted with the alkaline myofibrillar ATPase when there is a marked predominance of one fiber type.

NADH-Tetrazolium Reductase

This blue-black stain couples the reduction of tetrazolium with the oxidation of β-NADH. This oxidative stain defines mitochondria, the sarcoplasmic reticulum, and transverse tubules. Fibers show a finely reticulated pattern in which the type I fibers are dark and type II fibers are light. Gradations of the staining of type II fibers may be evident with IIb fibers very light and IIa fibers darker. Disturbances of the myofiber architecture are most readily demonstrated with this stain.

Succinic Dehydrogenase

This stain is of a purely mitochondrial enzyme and it demonstrates the distribution of mitochondria in fibers. The type I fibers with more mitochondria are dark, while type II fibers are light. Abnormalities of myofiber architecture disturb the normal distribution of mitochondria and can be demonstrated with this stain. Ragged-red fibers which contain an accumulation of bizarre mitochondria are especially evident with this stain.

Nonspecific Esterase

This stain shows endplates dark brown due to the high concentration of acetylcholinesterase. Age pigment (lipofuscin) demonstrates high esterase activity and appears as subsarcolemmal dark dots. Macrophages stain darkly due to the high esterase activity of activated lysosomes. Also, the lysosomes in lysosomal storage diseases (such as acid maltase deficiency) show a high level of esterase staining. Denervated fibers appear very dark.

Alkaline Phosphatase

In normal muscle specimens there may be minimal staining in the walls of large blood vessels, but often no staining can be demonstrated in the normal specimen. In myopathic disorders, the damaged degenerating fibers are easily demonstrated by staining black. In addition, newly synthesized connective tissue will show as streaks of black staining.

Menadione-linked α-Glycerophosphate Dehydrogenase

This blue stain of the intermyofibrillar aqueous phase is light in type I fibers and dark in type II fibers. It is useful in defining architectural disturbances in the type II fibers.

Periodic Acid–Schiff

Periodic acid–Schiff (PAS) stain will stain glycogen, glycoproteins, and neutral mucopolysaccharides a magenta color; nuclei are counterstained blue. The major PAS positive material in muscle fibers is the stored glycogen, which is greater in type II fibers than in type I fibers. Increased amounts of PAS staining glycogen can be demonstrated in glycogen storage diseases, such as myophosphorylase deficiency and acid maltase deficiency. After diastase digestion of the glycogen, the magenta PAS staining of glycoproteins and polysaccharides in blood vessel walls is evident. This is particularly useful in demonstrating the thickened capillaries in diabetes mellitus.

Alizarin Red S

This stain demonstrates calcium. Recently damaged fibers accumulate calcium and show excessive staining. In addition, there may be accumulation of calcium in connective tissue in inflammatory diseases of muscle.

Adenylate Deaminase

The common muscle disorder adenylate deaminase deficiency, characterized by muscle fatigue and exercise-induced muscle pain, can be demonstrated by absence of normal staining by this reaction.

Toluidine Blue

This stain can be used to demonstrate metachromatic staining materials (such as amyloid) or the abnormal accumulations in metachromatic leukodystrophy.

Oil Red O

With this stain, triglycerides and neutral fats appear as red droplets (nuclei are blue). Type I fibers normally contain more oil red O positive material (triglycerides) than type II fibers. Premenopausal females show greater fiber staining than males. Disorders of lipid metabolism in which such triglycerides are stored, such as carnitine deficiency, show excess oil red O staining.

Sudan Black B

This is an alternative stain for triglycerides and neutral fats, which appear blue-black (nuclei are blue). Sudan black B will also stain more complex lipids such as phospholipids and glycolipids.

Acid Phosphatase

This stain demonstrates lysosomes against a green background. In normal muscle, the reddish lysosomal staining is ordinarily too faint to detect, but when lysosomes become enlarged (as in activated macrophages or in lysosomal storage diseases, such as acid maltase deficiency) the red staining is very prominent. In addition, age pigment (lipofuscin) will appear as subsarcolemmal red dots.

Lactate Dehydrogenase

The presence of this enzyme in muscle fibers may be demonstrated by this blue stain.

NADPH-Tetrazolium Reductase

This blue-black stain couples the reduction of tetrazolium to the oxidation of NADPH. The staining pattern closely resembles NADH-tetrazolium reductase, but with a greater distinction between light type II and dark type I fibers.

Glyceraldehyde-3-Phosphate Dehydrogenase

This stain demonstrates the presence of this glycolytic enzyme in muscle fibers. No disorder of glyceraldehyde-3-phosphate dehydrogenase (GR-3-PD) deficiency has yet been reported, but an animal model with GR-3-PD deficiency mimics the symptomatology of humans with either myophosphorylase deficiency or phosphofructokinase deficiency.

Amylophosphorylase

The stain for amylophosphorylase bound to glycogen utilizes the interaction of iodine with glycogen molecules of varying chain length resulting in colors from dark blue-black for longest chains through purple to light red for short chains. The type II muscle fibers show the darkest color with this stain, while the type I fibers have light staining. The major usefulness of this stain is in demonstrating the absence of muscle amylophosphorylase in McArdle's disease (myophosphorylase deficiency) which is characterized by exercise-induced painful muscle cramps and postexercise rhabdomyolysis.

Adenyl Cyclase

Adenyl cyclase stains large and small blood vessel walls, including capillaries. It is particularly useful in evaluating abnormalities in vascularity, such as may occur in inflammatory diseases of muscle or diabetes mellitus.

Guanidyl Cyclase

Guanidyl cyclase technique stains the large blood vessel walls and can be used in conjunction with the adenyl cyclase stain to evaluate vascularity.

Wright's Stain

This is the stain commonly used to assess blood smears and bone marrow preparations. It is most useful in assessing the types of inflammatory cells present in an inflammatory disease of muscle. The presence of eosinophils would suggest a parasitic disease, neutrophils would suggest acute inflammation, and monocytes and lymphocytes would suggest a chronic inflammatory disorder.

3 LIGHT MICROSCOPY OF HISTOCHEMICALLY STAINED MUSCLE BIOPSY SPECIMENS

Muscle biopsy specimens stained histochemically (as described in chapter 2) reveal a wide variety of colors when viewed through the microscope. Color changes or changes in the distribution of specific colors are important diagnostic characteristics. Unfortunately, black and white photomicrographs (used in most publications) obscure these necessary histological details. The following color photomicrographs are representative examples of histochemically stained normal and diseased muscle specimens and the salient features of each photomicrograph are described in the legends. Plates 1 through 14 are photomicrographs of normal human muscle. Characteristic histochemical staining patterns of common and rare neuromuscular disorders are shown in Plates 15 through 63. Some of the troublesome common artifacts are demonstrated in Plates 64 through 67. The rat and guinea pig muscle specimens shown in Plates 68 and 69 have normal histological features that would be considered abnormal if observed in human muscle biopsies; familiarity with the full range of normal (for that species) is necessary before interpreting nonhuman muscle specimens.

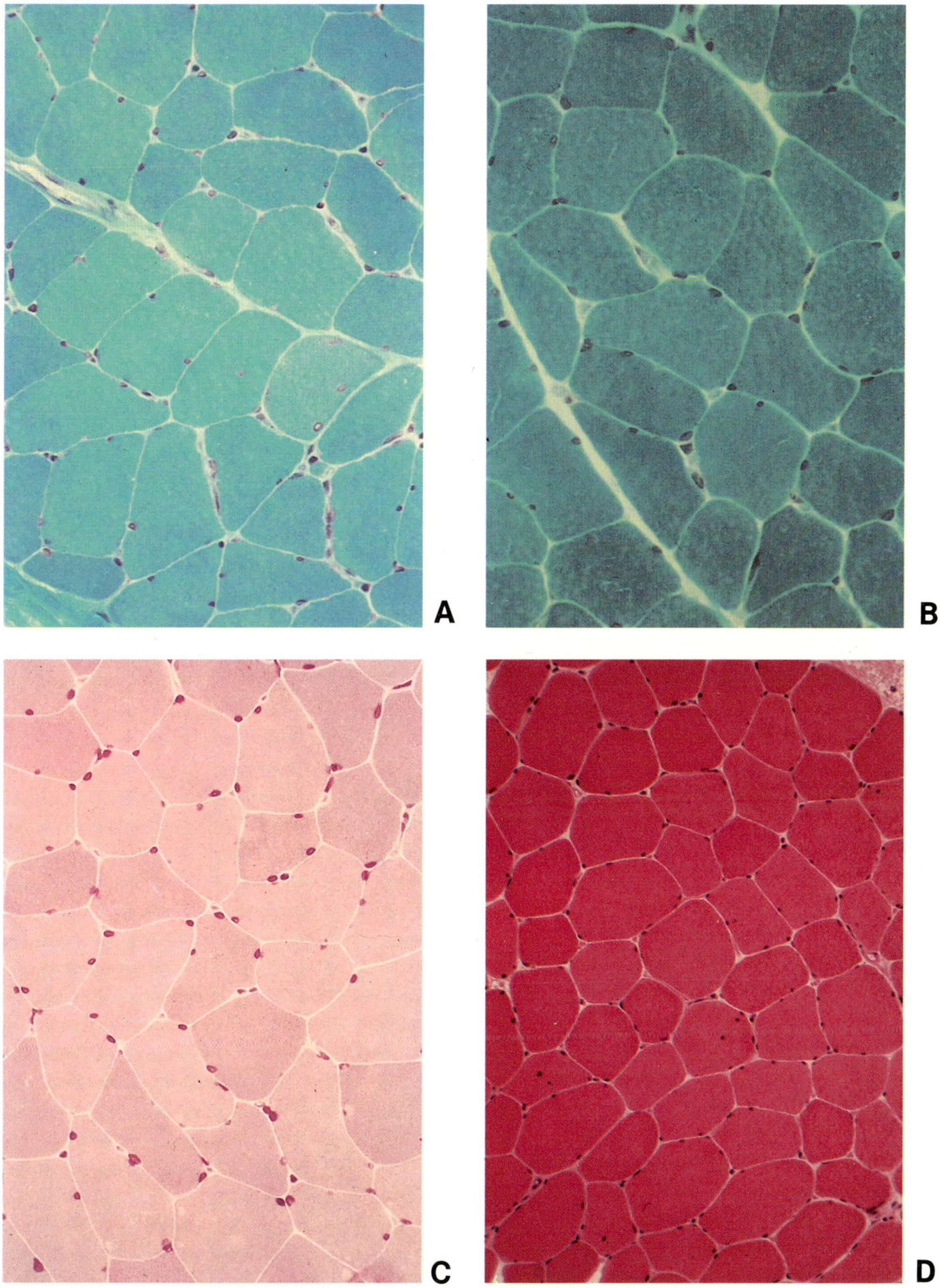

Plate 1 Cross sections of normal muscle. **A,B:** Modified Gomori trichrome stain. **C,D:** H&E stain. Note that even though the fibers are cylindrical, not all are sectioned perfectly across and thus have irregular shapes; however, the maximum diameter across the lesser aspect of all the muscle fibers is essentially uniform. The average diameter of muscle fibers in adults is approximately 50 μm. Some differentiation of two fiber types can occasionally be seen with the modified Gomori trichrome and H&E stains (type I fibers darker than type II fibers). The muscle fibers are multinucleate cells with the nuclei peripherally placed. Connective tissue stains green with the modified Gomori trichrome stain and there is very little endomysial connective tissue between individual muscle fibers.

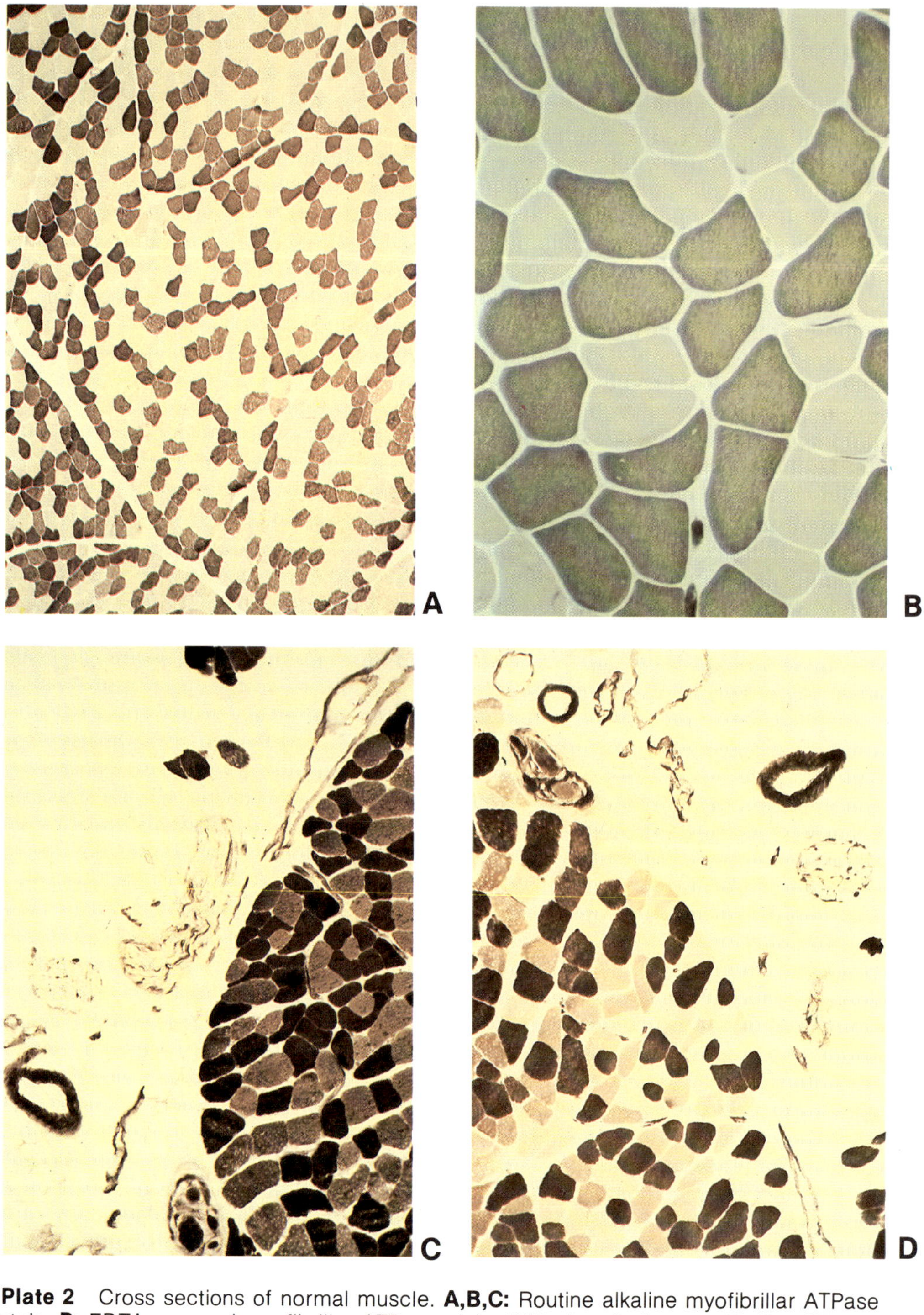

Plate 2 Cross sections of normal muscle. **A,B,C:** Routine alkaline myofibrillar ATPase stain. **D:** EDTA-reversed myofibrillar ATPase stain. With the alkaline myofibrillar ATPase stain, the individually stained myofibrils are separated by an unstained intermyofibrillar network. With this stain, two distinct muscle fiber types are identifiable—type I fibers are light, while type II fibers are dark. There is a fairly even checkerboard distribution of the two muscle fiber types. The intrafusal muscle fibers of muscle spindles and the smooth muscle of blood vessels also stain dark (**C**). With the EDTA-reversed myofibrillar ATPase stain, the type I fibers stain dark, while the type II fibers stain light.

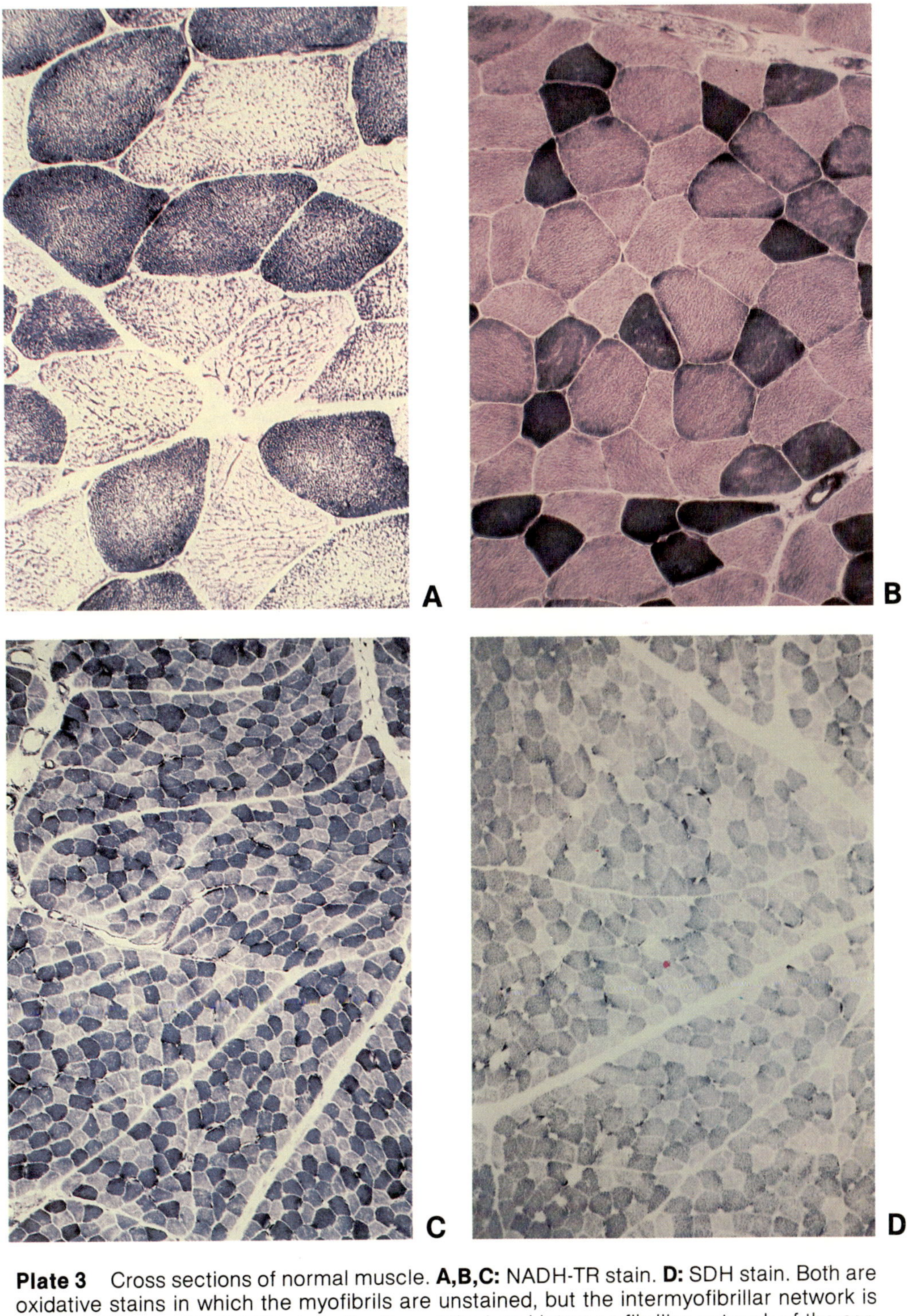

Plate 3 Cross sections of normal muscle. **A,B,C:** NADH-TR stain. **D:** SDH stain. Both are oxidative stains in which the myofibrils are unstained, but the intermyofibrillar network is stained. With the NADH-TR stain, the fine, reticulated intermyofibrillar network of the sarcoplasmic reticulum and the peripherally positioned mitochondria are stained. The SDH stain stains only the mitochondria in a dot-like pattern. The ATPase type I fibers are dark and the type II fibers are light with both the NADH-TR and SDH stains. There is often an area of increased subsarcolemmal staining adjacent to capillaries.

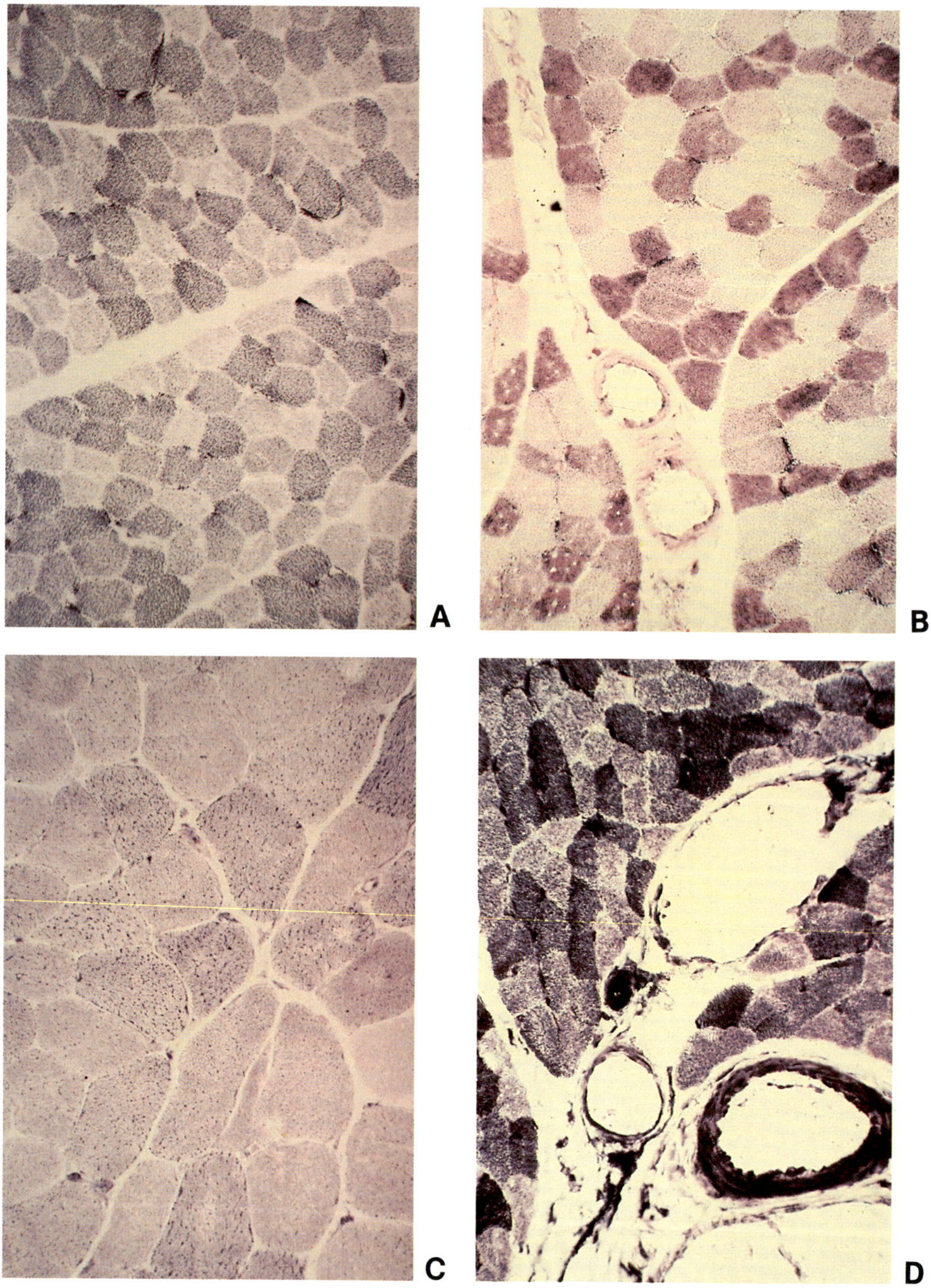

Plate 4 Cross sections of normal muscle. **A:** SDH stain. **B:** NADPH-TR stain. **C:** Glyceraldehyde-3-phosphate dehydrogenase stain. **D:** Lactic dehydrogenase stain. There are several tetrazolium reductase-linked staining techniques which stain ATPase type I fibers dark and type II fibers light. The NADPH-TR stain shows a more striking light-dark differentiation of fiber types, but with a less clear definition of the intermyofibrillar network than does the NADH-TR. The presence of enzymatic activity of the glycolytic enzymes glyceraldehyde-3-phosphate dehydrogenase and lactic dehydrogenase may be demonstrated by histochemical staining.

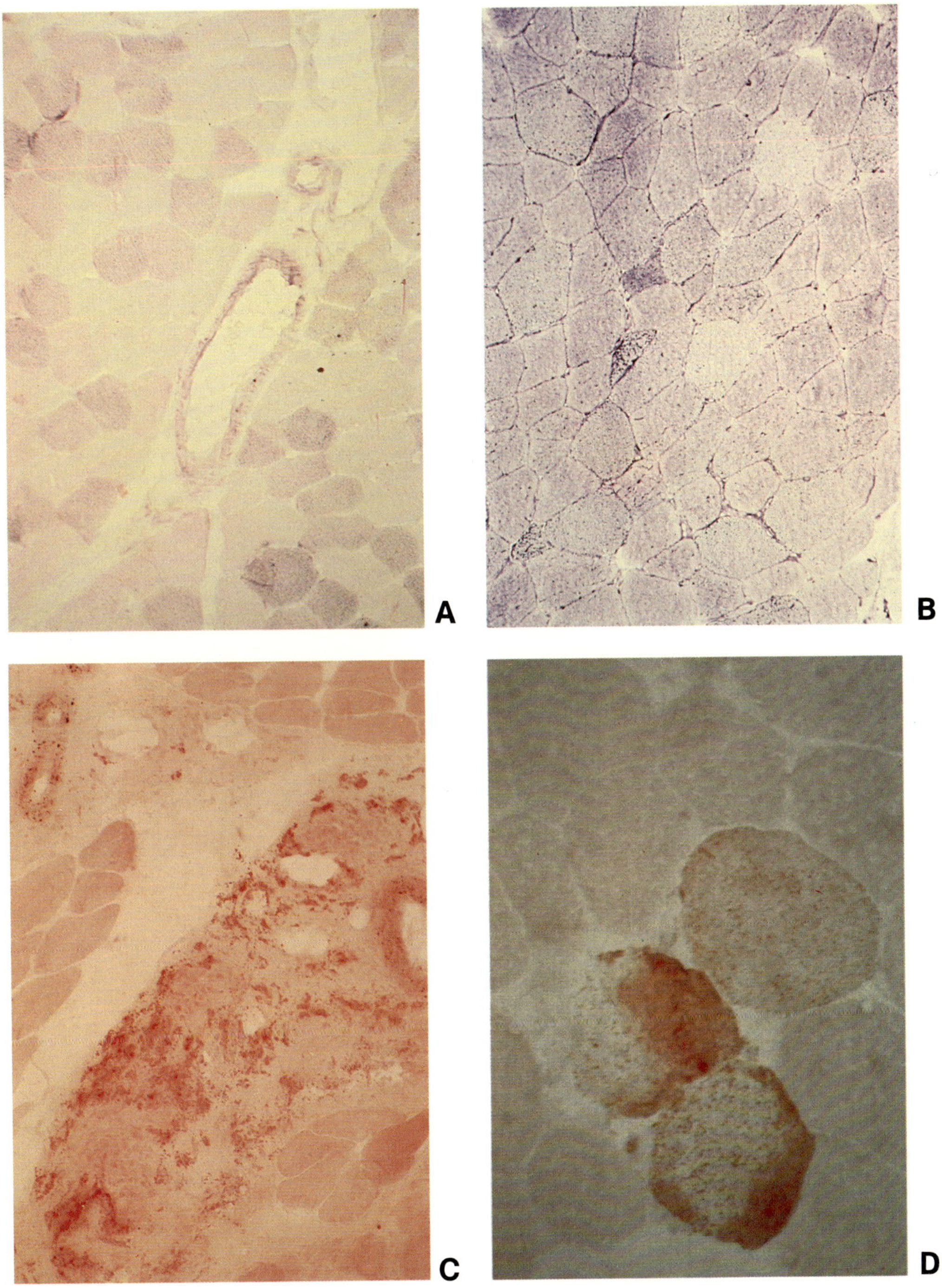

Plate 5 Cross sections of normal muscle (**A,B,C**) and cross section of muscle with three damaged muscle fibers (**D**). **A:** Menadione-linked alpha-glycerophosphate dehydrogenase. **B:** Adenylate deaminase. **C,D:** Alizarin red S. Both the menadione-linked alpha-glycerophosphate dehydrogenase and the adenylate deaminase stains are tetrazolium reductase-linked staining techniques. However, while demonstrating the intermyofibrillar network, the menadione-linked alpha-glycerophosphate dehydrogenase stain shows the ATPase type II fibers dark and the ATPase type I fibers light. The alizarin red S stain for calcium ordinarily shows the calcium as a faint homogeneous red staining of individual fibers (**C**). Damaged fibers in which there is excessive calcium accumulation (**D**) will show various patterns of excessive staining. Calcium staining may also be seen in connective tissue and blood vessel walls as a normal part of aging (as in the central portion of **C**).

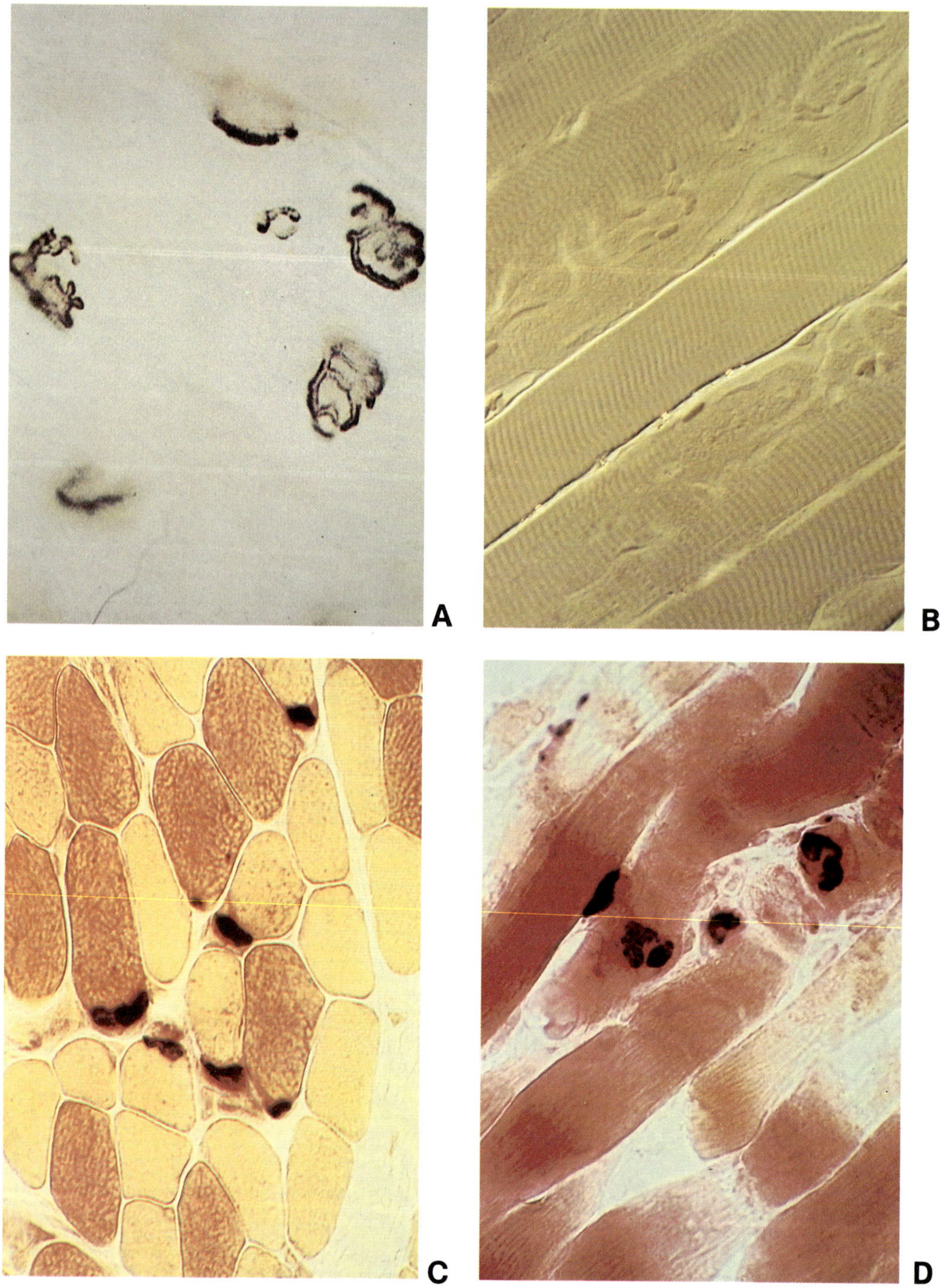

Plate 6 Cross sections (**A,C**) and longitudinal sections (**B,D**) of normal muscle. **A,B:** Unstained muscle with Nomarski interference optics. **C,D:** Nonspecific esterase stain. The neuromuscular junction may occasionally be seen in sections of muscle specimens taken near the motor point (site of nerve entry). The neuromuscular junction is a tortuous, infolded structure which stains dark with the nonspecific esterase stain because of the high concentration of acetylcholinesterase activity.

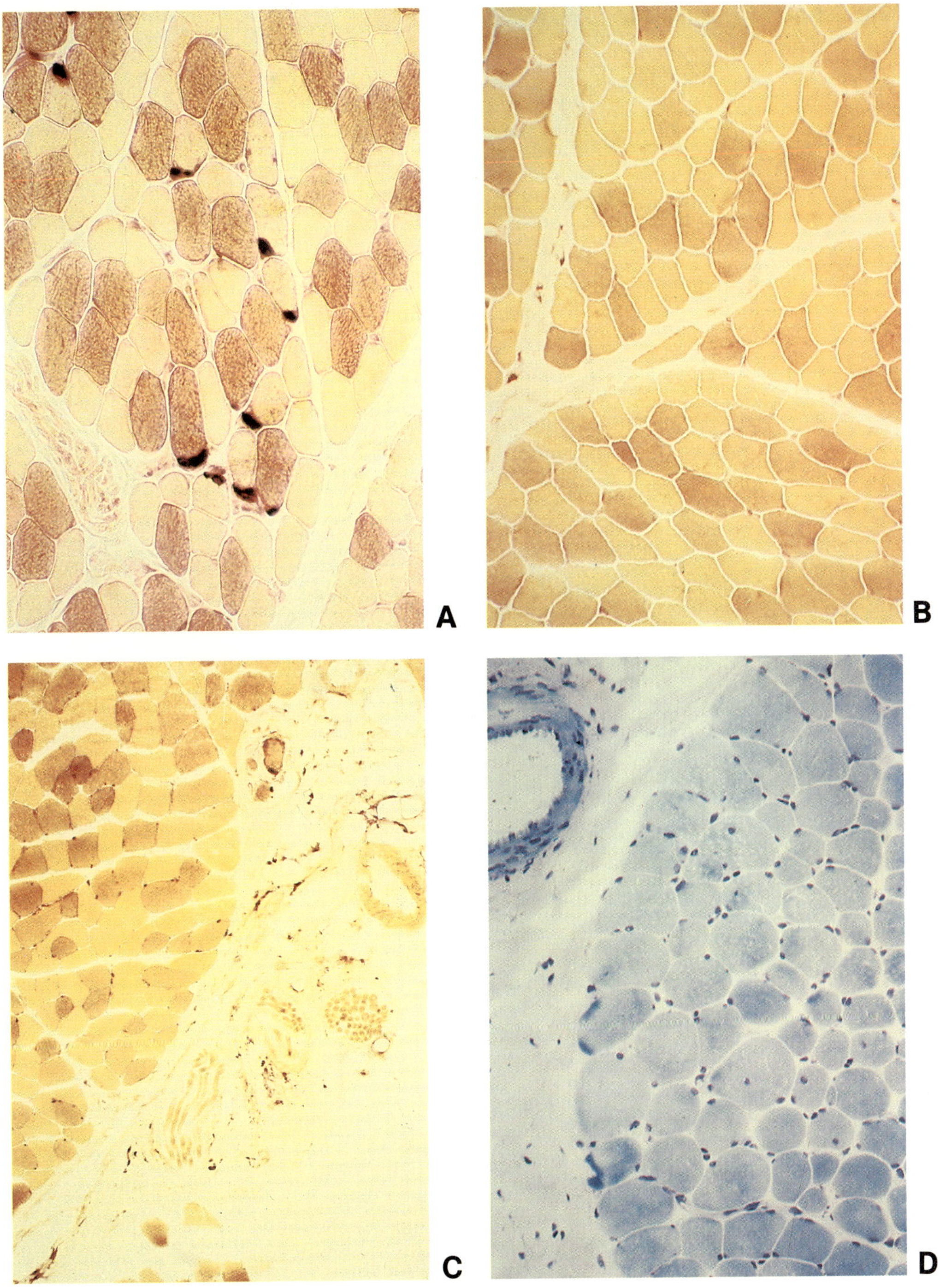

Plate 7 Cross sections of normal muscle. **A,B,C:** Nonspecific esterase stain. **D:** Toluidine blue stain. The nonspecific esterase stain demonstrates the neuromuscular junction and stains ATPase type I fibers dark and ATPase type II fibers light. The toluidine blue stain is a metachromatic stain which in normal muscle shows homogeneous blue coloration. The nuclei are well demonstrated with toluidine blue.

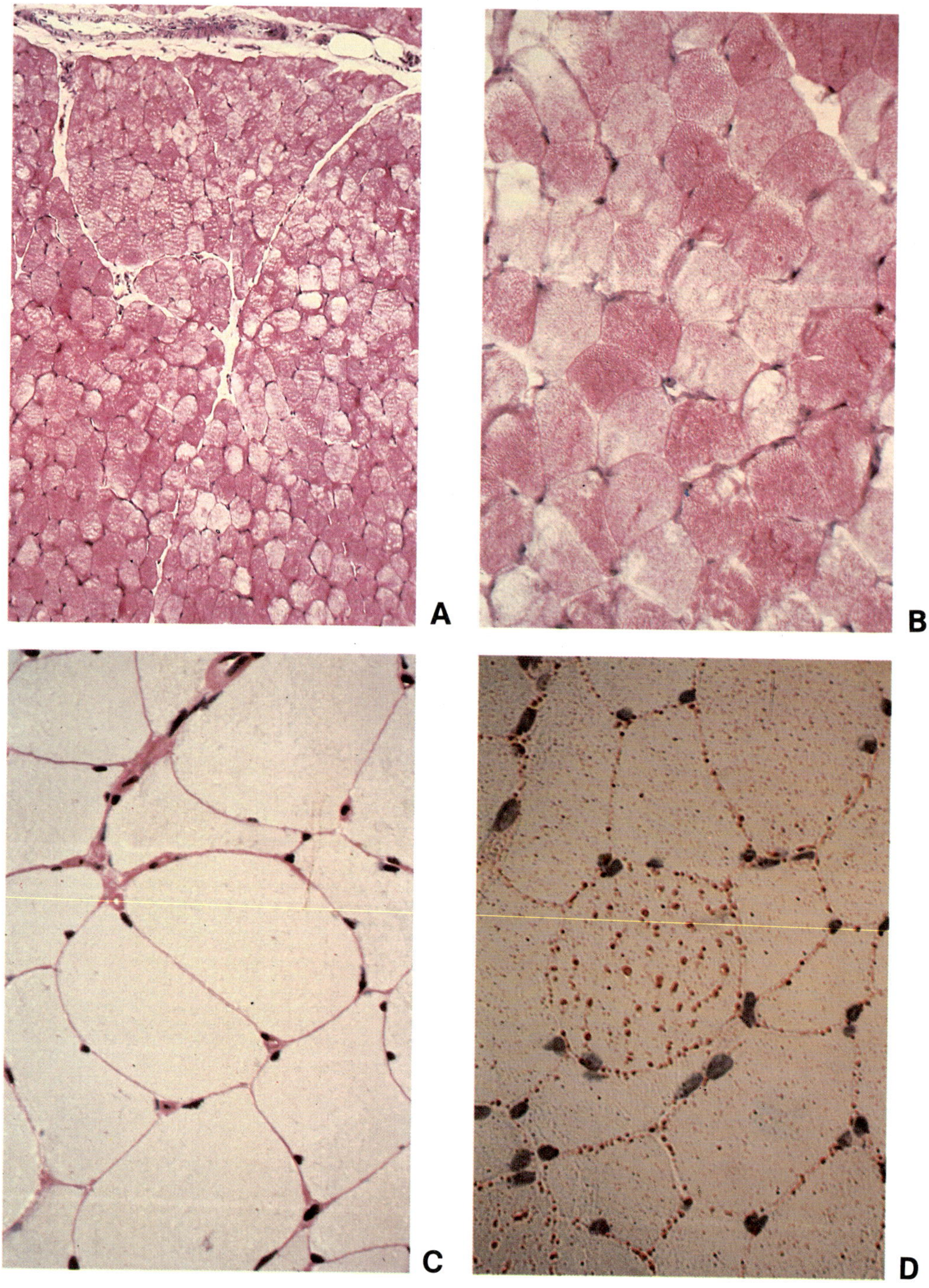

Plate 8 Cross sections of normal muscle. **A,B:** Periodic acid–Schiff (PAS). **C:** PAS after preincubation with diastase (negative control). **D:** Oil red O stain. The PAS stain demonstrates carbohydrate in the muscle specimen. Glycogen is the most abundant carbohydrate, and ATPase type II fibers, with their larger content of glycogen, stain darker than ATPase type I fibers. After diastase digestion of the glycogen, only the glycoproteins of the cell membrane and blood vessel walls are stained. The oil red O lipid stain demonstrates fats, principally triglycerides and other neutral fats, as small reddish droplets in the fibers. Staining is more highly concentrated in the ATPase type I muscle fibers.

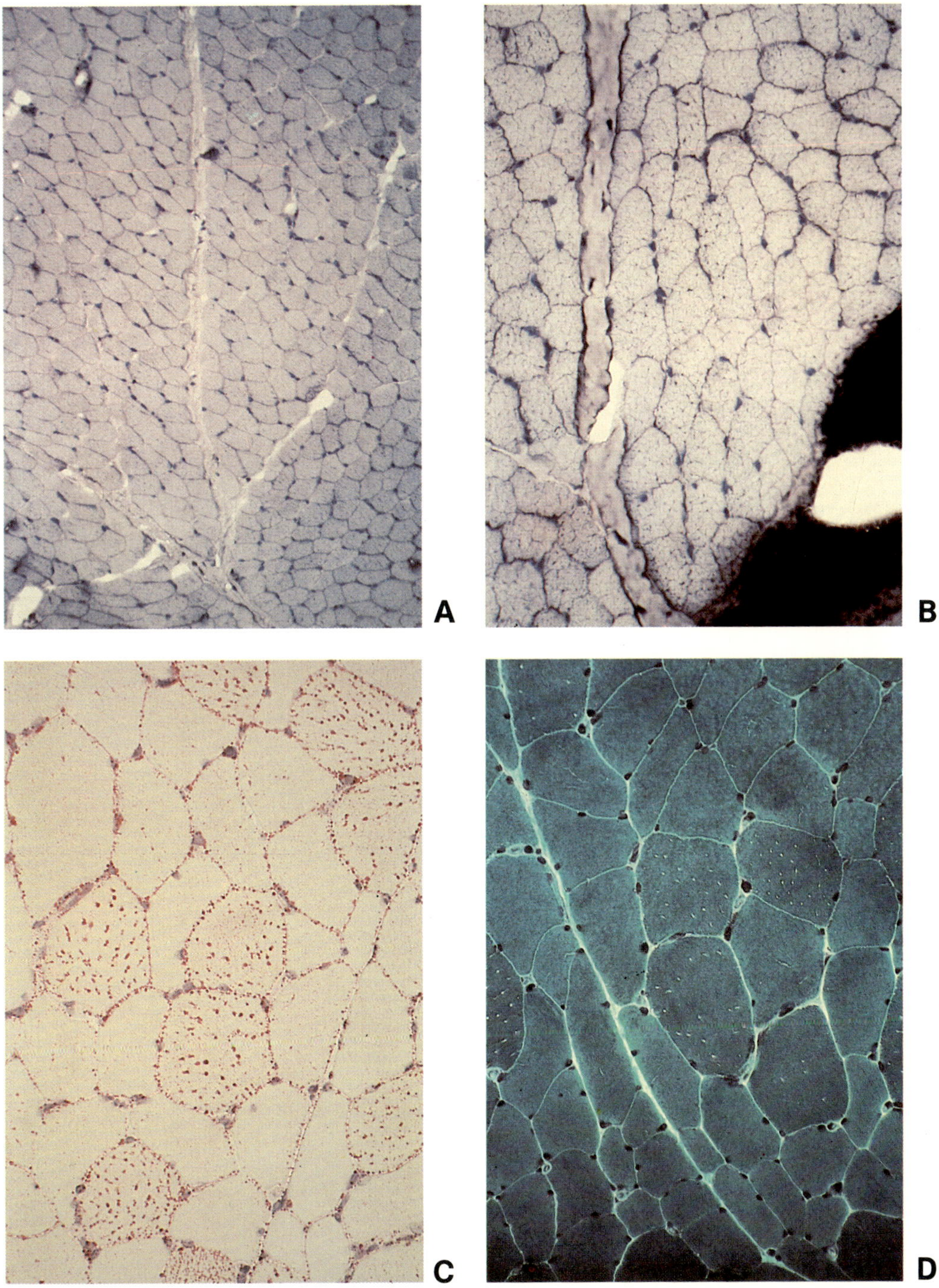

Plate 9 Cross sections of normal muscle. **A,B:** Sudan black B stain. **C:** Oil red O stain. **D:** Modified Gomori trichrome stain. Lipids may be stained with either oil red O or Sudan black B. While the oil red O stain tends to show more of the simple lipids such as triglycerides and neutral fats, the Sudan black B also stains the complex lipids such as phospholipids and glycolipids. The dark staining of lipid deposits in the connective tissue is evident in **B**. Occasionally, small unstained "holes" in the type I fibers (such as in the modified Gomori trichrome stain, **D**) will stain dark for lipid with the oil red O stain. This may be normal, especially in females. Compare **C** and **D**.

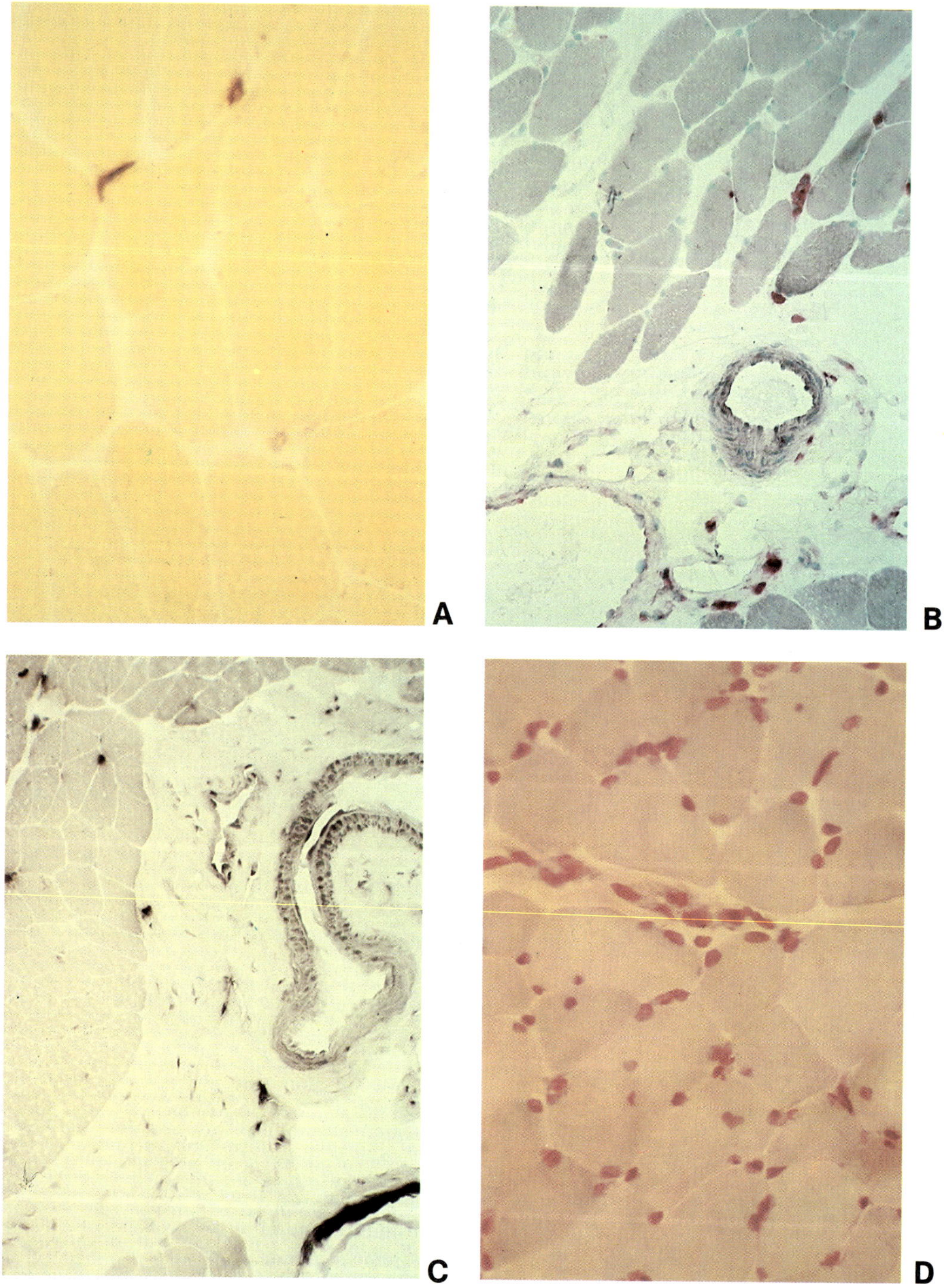

Plate 10 Cross sections of normal muscle. **A:** Alkaline phosphatase stain. **B:** Acid phosphatase stain. **C:** Guanidyl cyclase stain. **D:** Wright's stain. The alkaline phosphatase stain normally shows no fiber staining, but occasional minor deposits near blood vessels. There is normally no acid phosphatase staining (reddish) in fibers, with only the green counterstain evident. The guanidyl cyclase stain shows the individual endothelial cells of large blood vessels. Wright's stain demonstrates cell nuclei clearly.

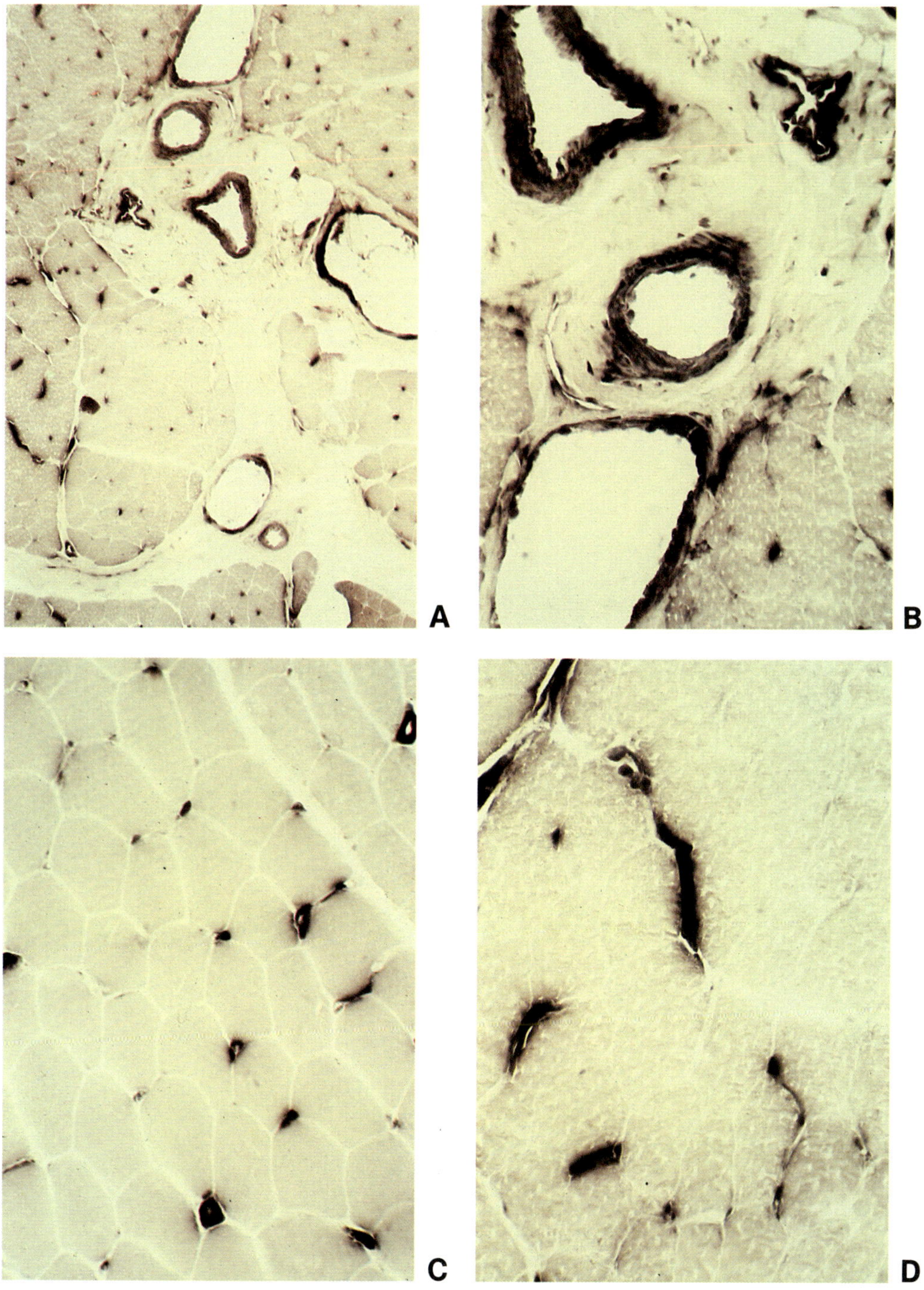

Plate 11 Cross sections of normal muscle. **A,B,C,D:** Adenyl cyclase stain. This staining technique clearly shows blood vessels, both large vessels (**A,B**) and the small interfiber capillary system (**C,D**).

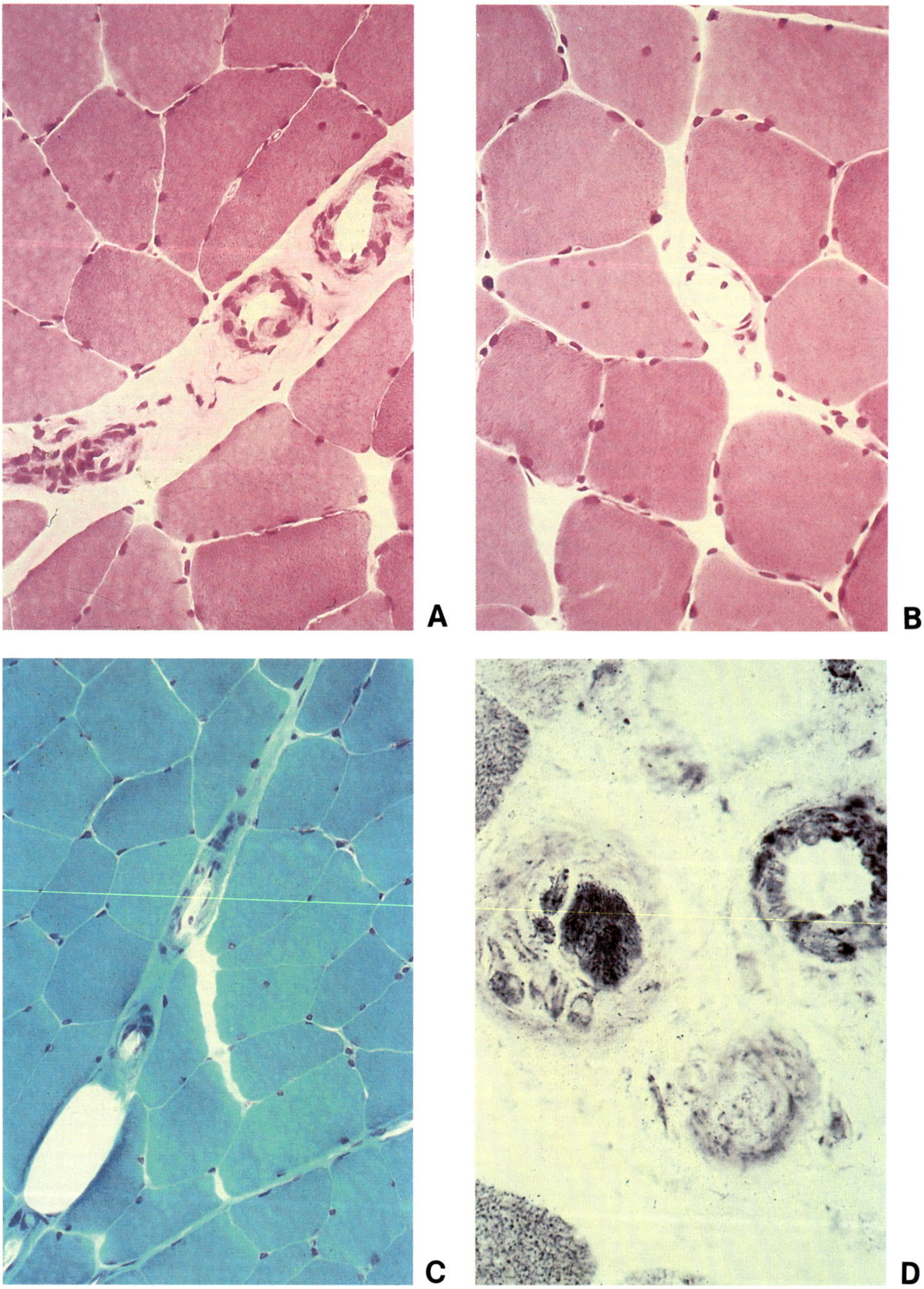

Plate 12 Cross sections of normal muscle. **A,B:** H&E stain. **C:** Modified Gomori trichrome. **D:** NADH-TR stain. There are usually multiple blood vessels evident in a muscle biopsy specimen. The vessels range in size from arterioles to small capillaries. Muscle spindles are also frequently evident (**D**).

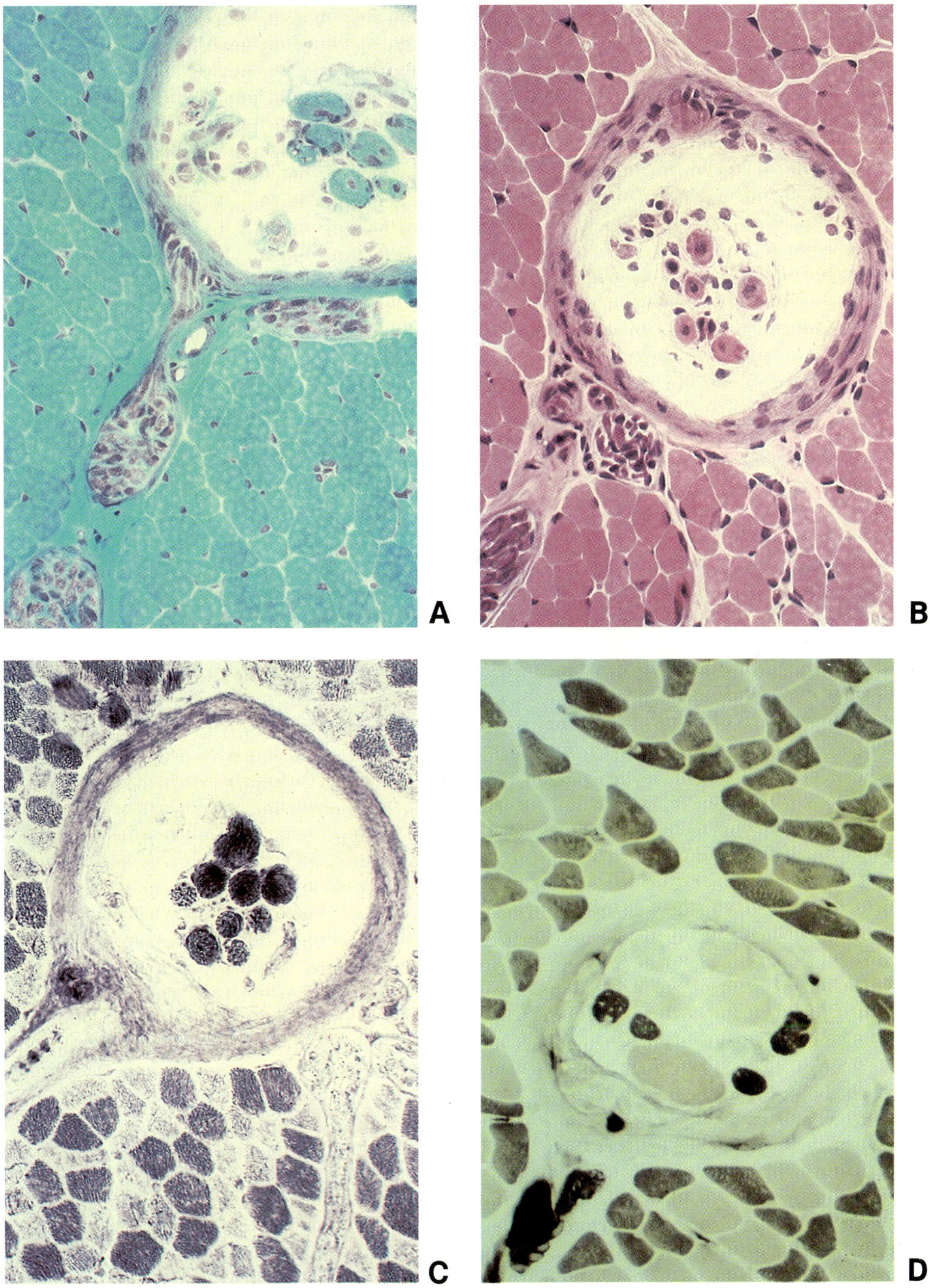

Plate 13 Cross sections of normal muscle showing muscle spindle at the site of entry of its specialized motor (gamma efferent) and sensory nerve supply. **A:** Modified Gomori trichrome stain. **B:** H&E stain. **C:** NADH-TR stain. **D:** Alkaline myofibrillar ATPase stain. Note that the muscle spindle can be identified from the surrounding muscle by its fibrous capsule. The intrafusal muscle fibers stain with the histochemical staining techniques. With the NADH-TR stain, the nuclear chain fibers appear homogeneously very dark, the nuclear bag_1 fibers have finely granular dark staining, and the nuclear bag_2 fibers have coarsely granular dark staining. With the alkaline myofibrillar ATPase stain, the nuclear chain fibers are dark and the nuclear bag fibers are light.

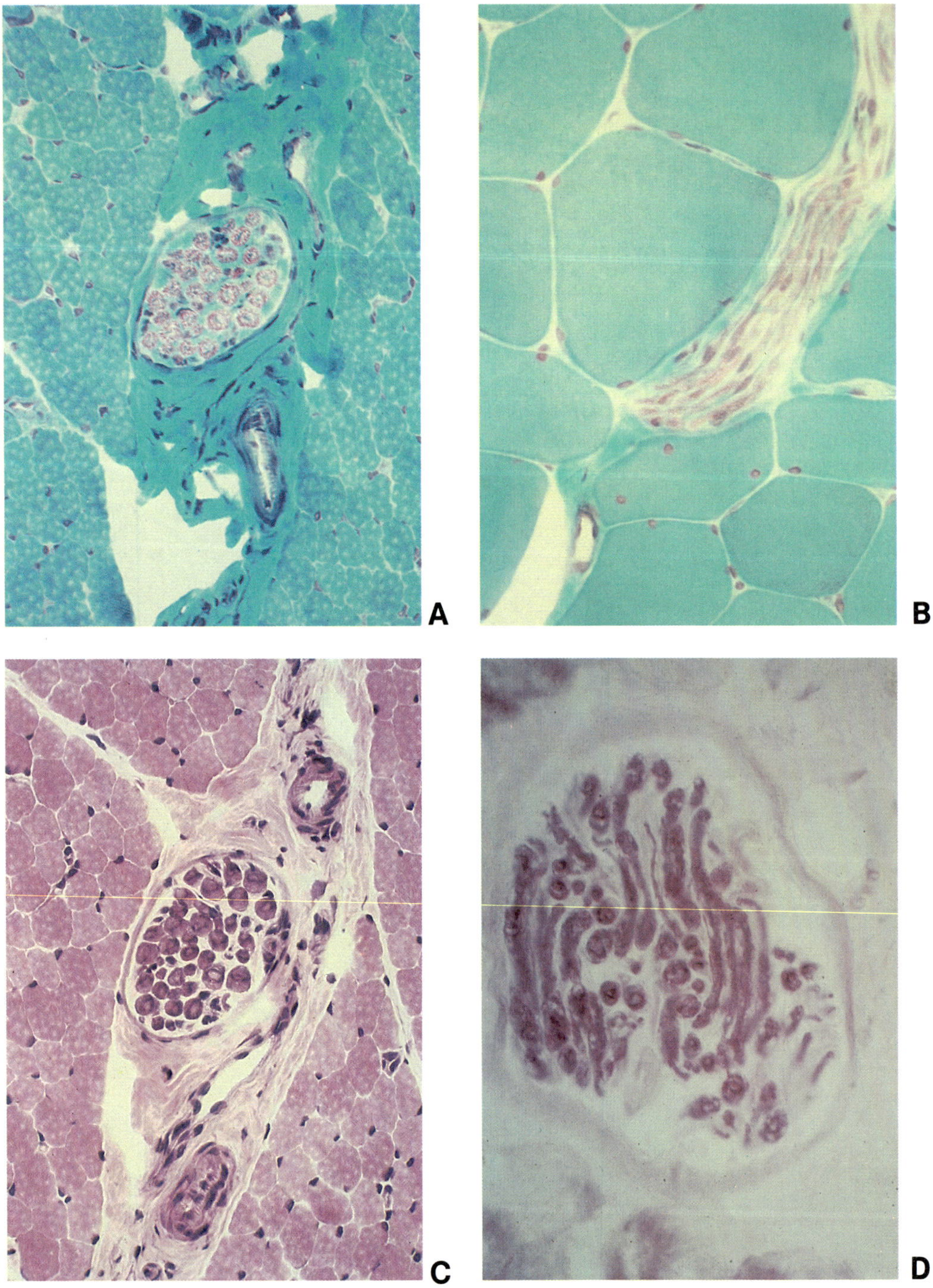

Plate 14 Cross sections of normal muscle showing intramuscular nerve twig. **A,B:** Modified Gomori trichrome stain. **C:** H&E stain. **D:** PAS stain. The myelin of the nerve fibers stains reddish with the modified Gomori trichrome stain and normally the nerve appears packed with fibers. The myelin also stains positive with the PAS stain.

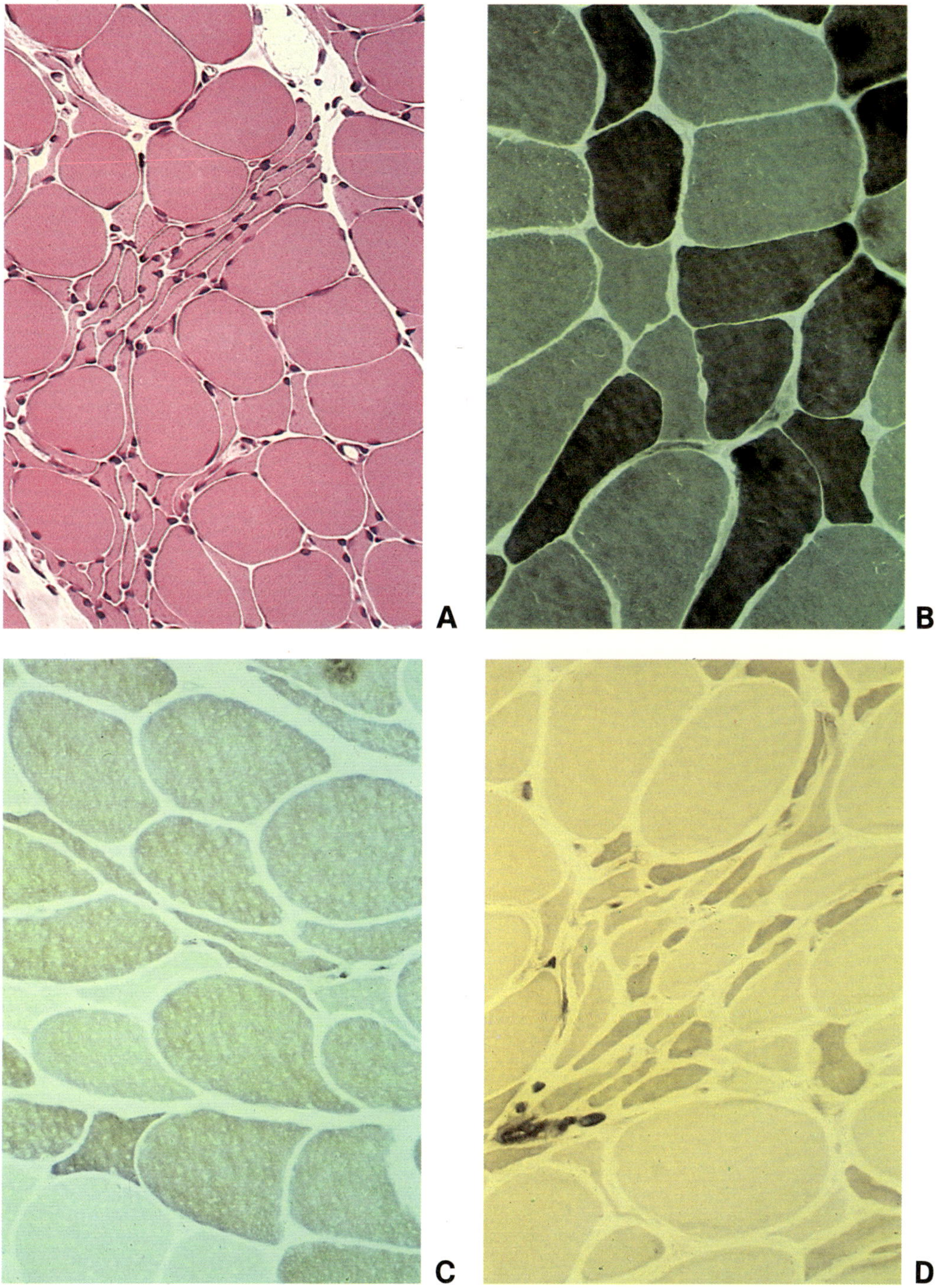

Plate 15 Cross sections of muscle showing changes of denervation. **A:** H&E stain. **B,C,D:** Alkaline myofibrillar ATPase stain. Note that the small fibers are both type I and type II with angulated corners (angular fibers) some of which are in groups (**A**).

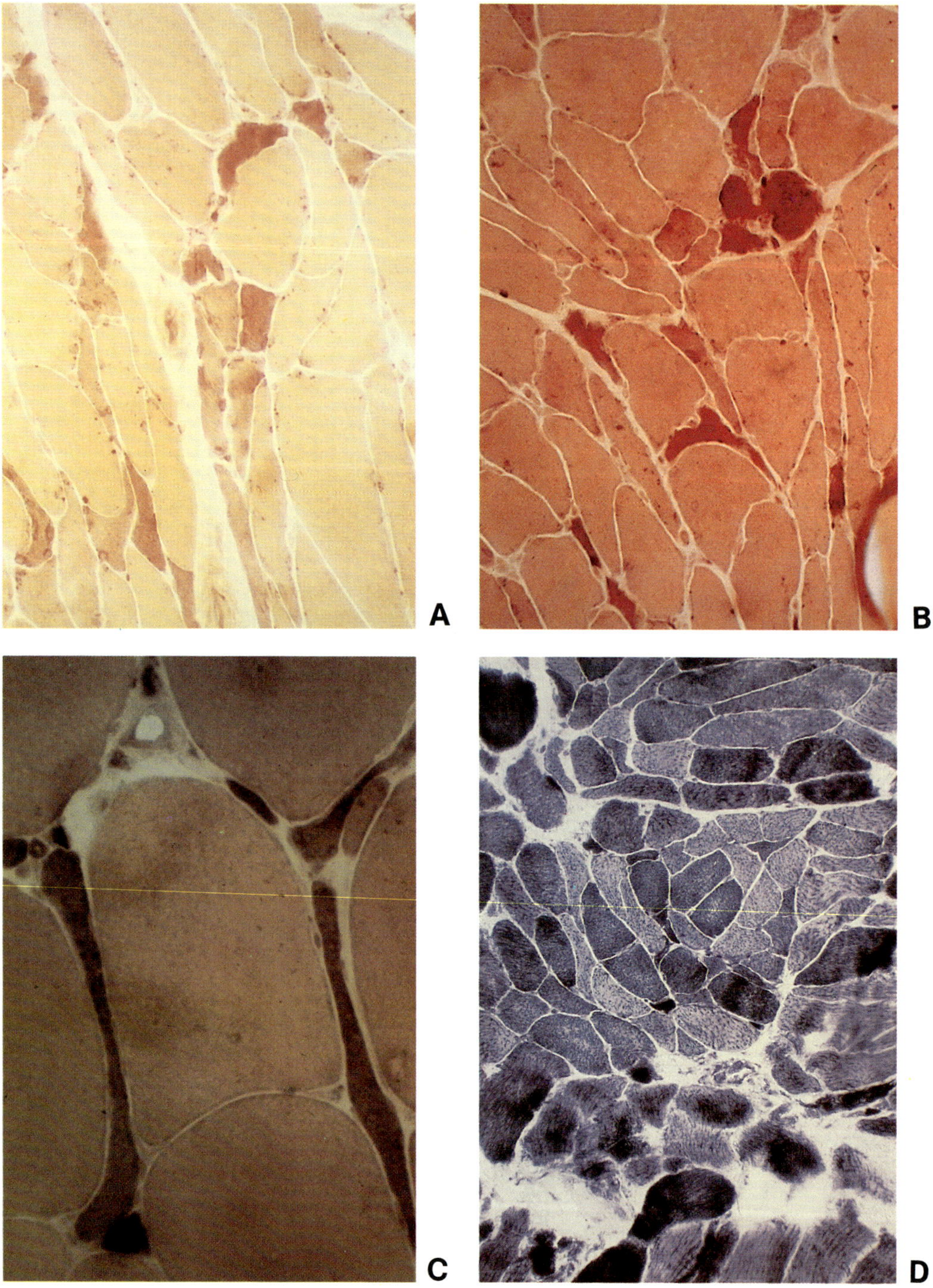

Plate 16 Cross sections of muscle showing denervation changes. **A,B,C:** Nonspecific esterase stain. **D:** NADH-TR stain. With both the nonspecific esterase stain and the NADH-TR stain, the small angular denervated fibers are excessively dark.

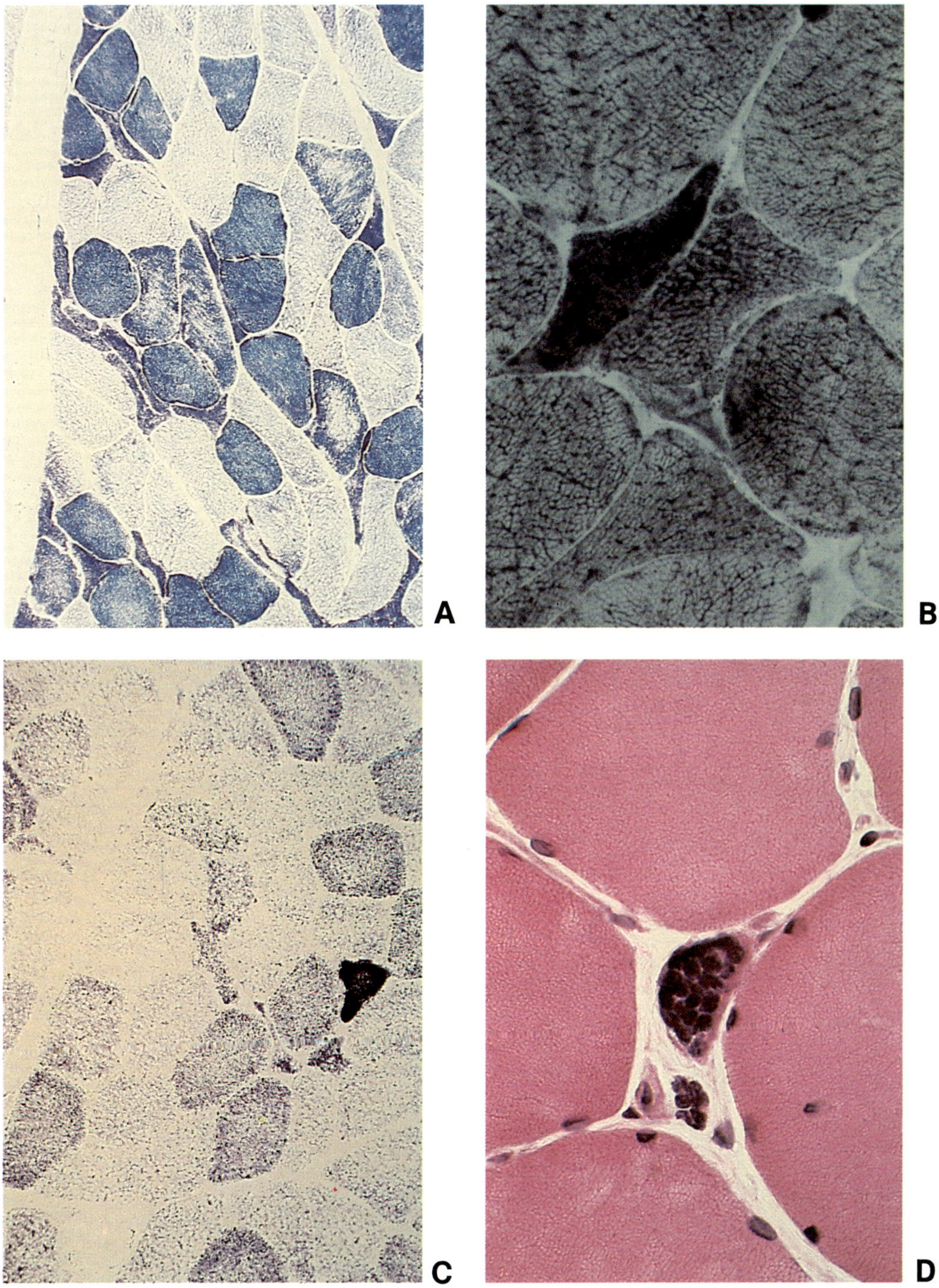

Plate 17 Cross sections of muscle showing denervation changes. **A,B:** NADH-TR stain. **C:** SDH stain. **D:** H&E stain. The small angular denervated fibers appear excessively dark with both NADH-TR and SDH stains. As a denervated fiber atrophies, only a remnant clump of pyknotic nuclei (**D**) may remain.

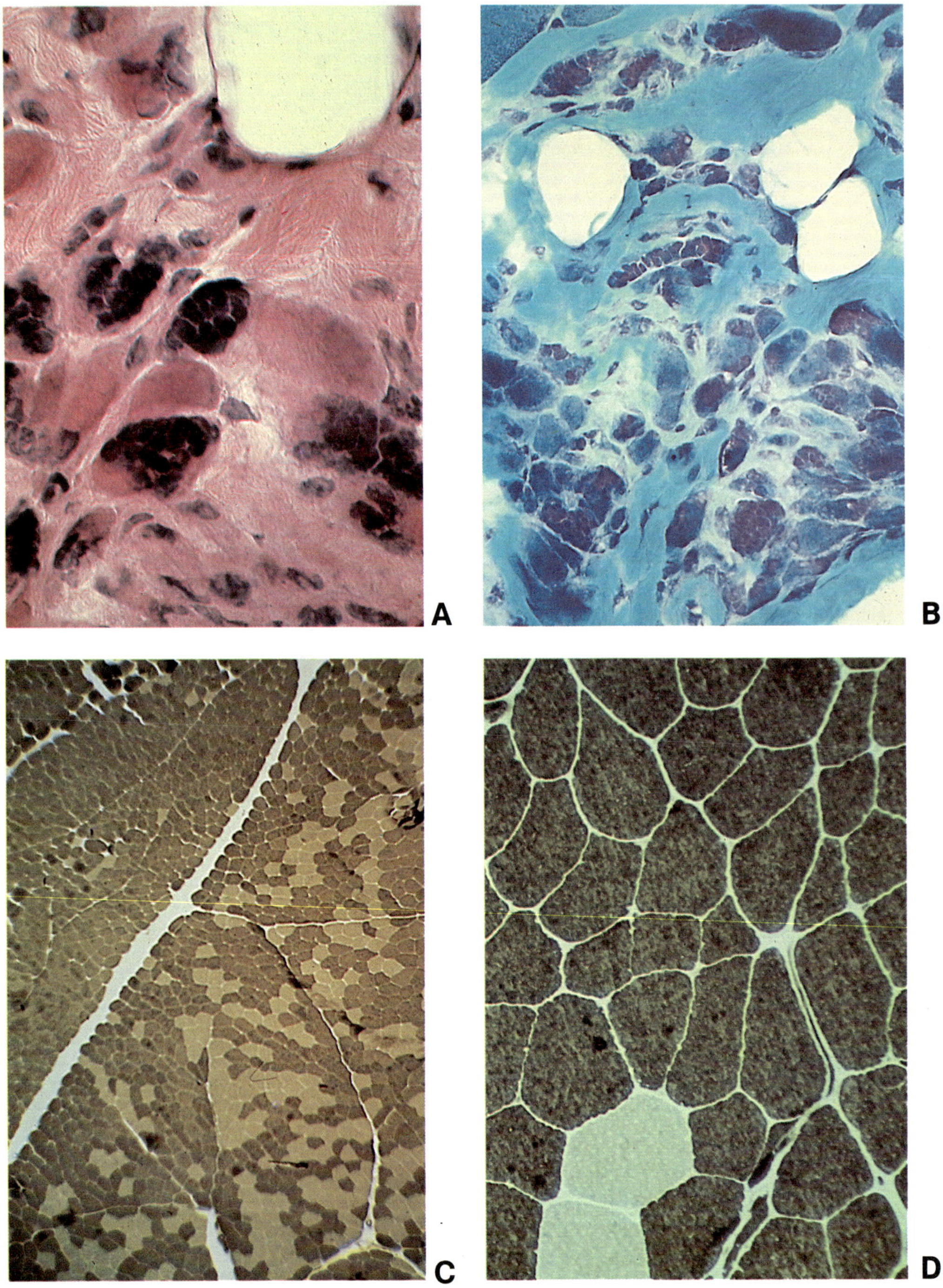

Plate 18 Cross sections of muscle showing denervation changes. **A:** H&E stain. **B:** Modified Gomori trichrome stain. **C,D:** Alkaline myofibrillar ATPase stain. Pyknotic nuclear clumps are the end result of long-standing denervation of muscle cells (**A,B**). With reinnervation of previously denervated fibers, there is loss of the usual checkerboard distribution of fiber types leading to large groups of a single fiber type (**C,D**).

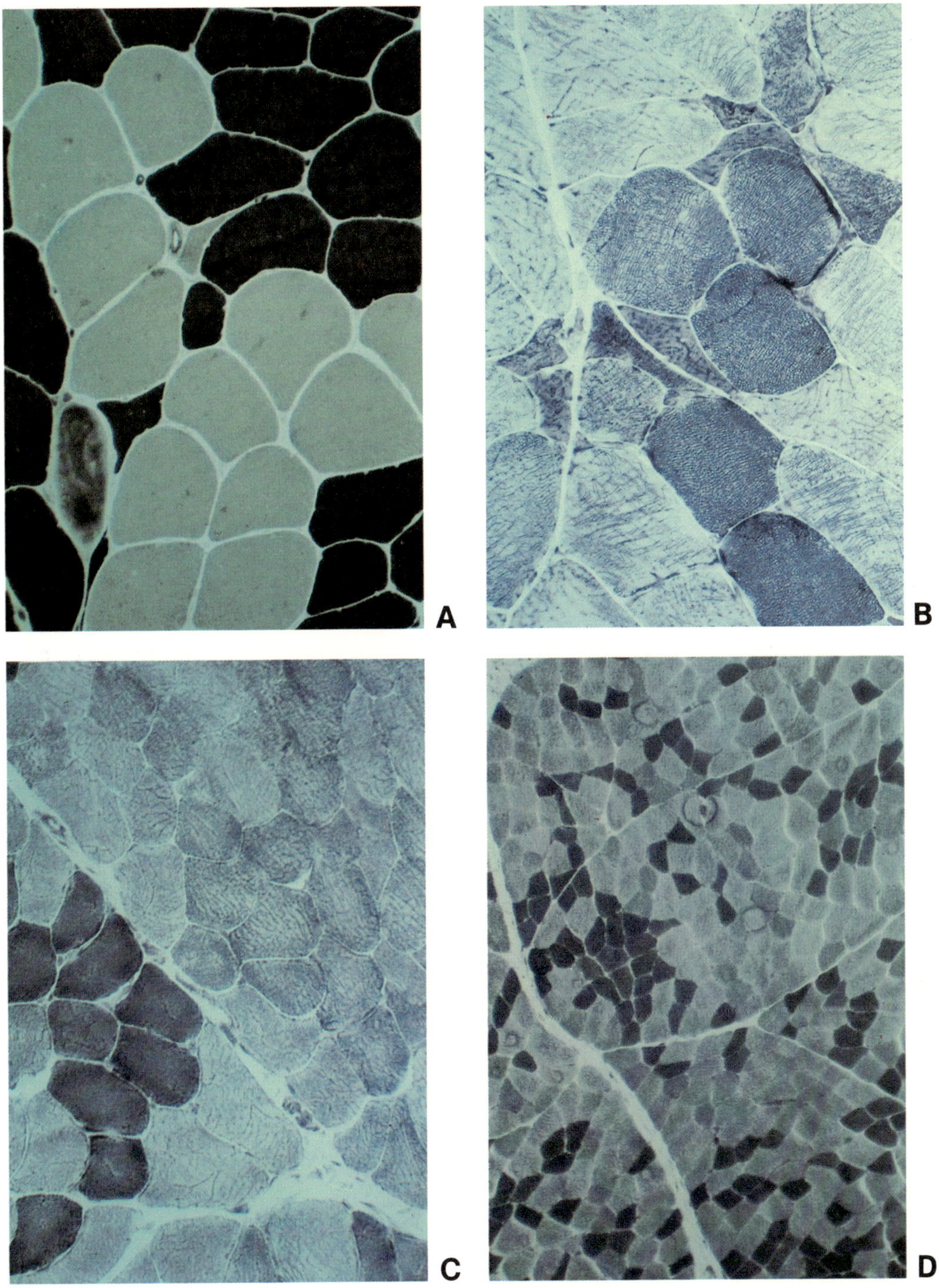

Plate 19 Cross sections of muscle showing denervation changes. **A:** Alkaline myofibrillar ATPase stain. **B,C,D:** NADH-TR stain. Atrophied angular fibers (**A,B**) and fiber type grouping (**C,D**) are usual features of chronic denervation.

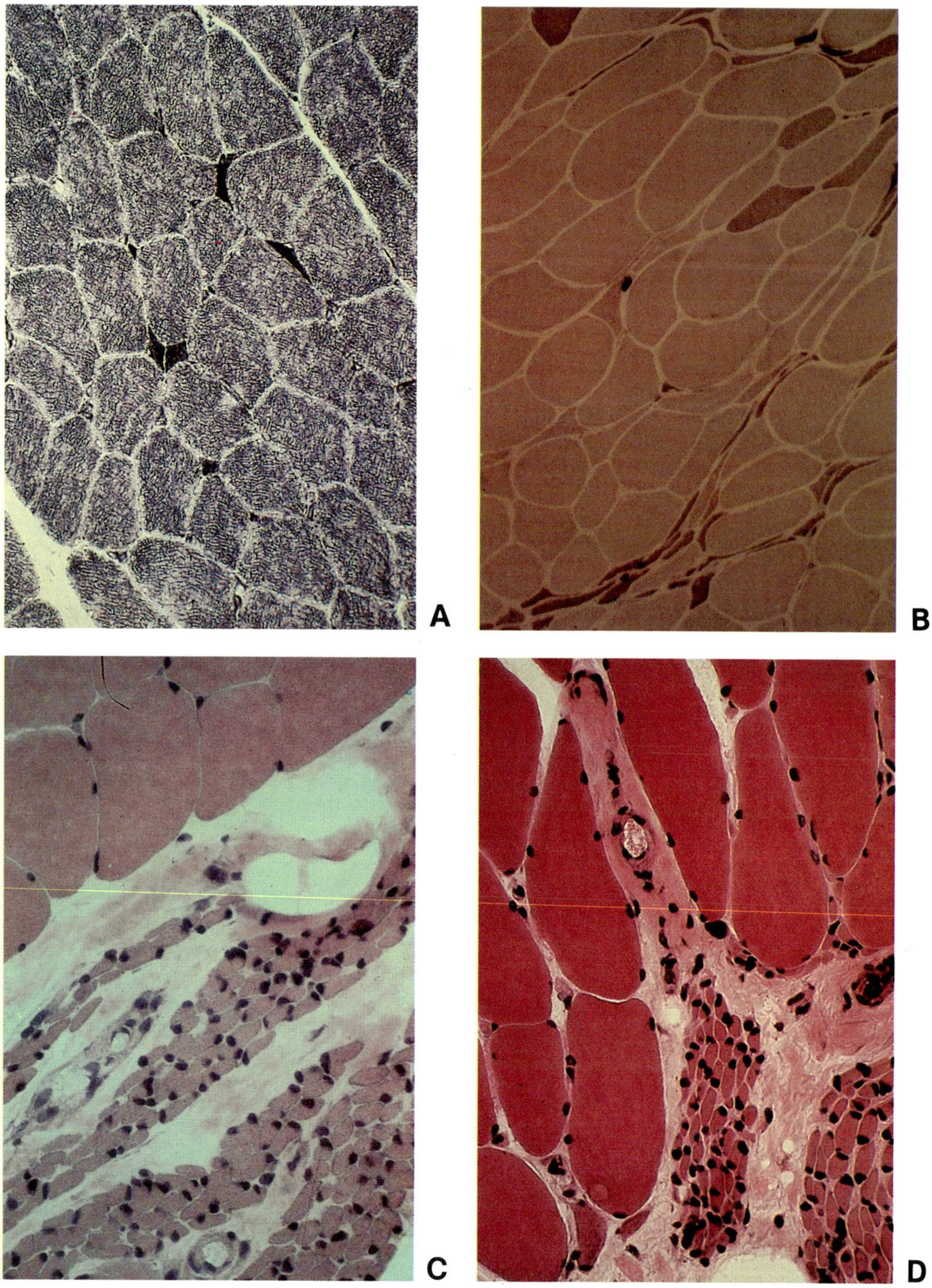

Plate 20 Cross sections of muscle showing denervation changes. **A:** NADH-TR stain. **B:** Alkaline myofibrillar ATPase stain. **C,D:** H&E stain. Denervation can produce either scattered angular atrophic fibers (**A,B**) or large groups of atrophic fibers (**C,D**) many of which will be more rounded. As in Plate **15D**, both type I and II fibers are atrophic and angular (**B**).

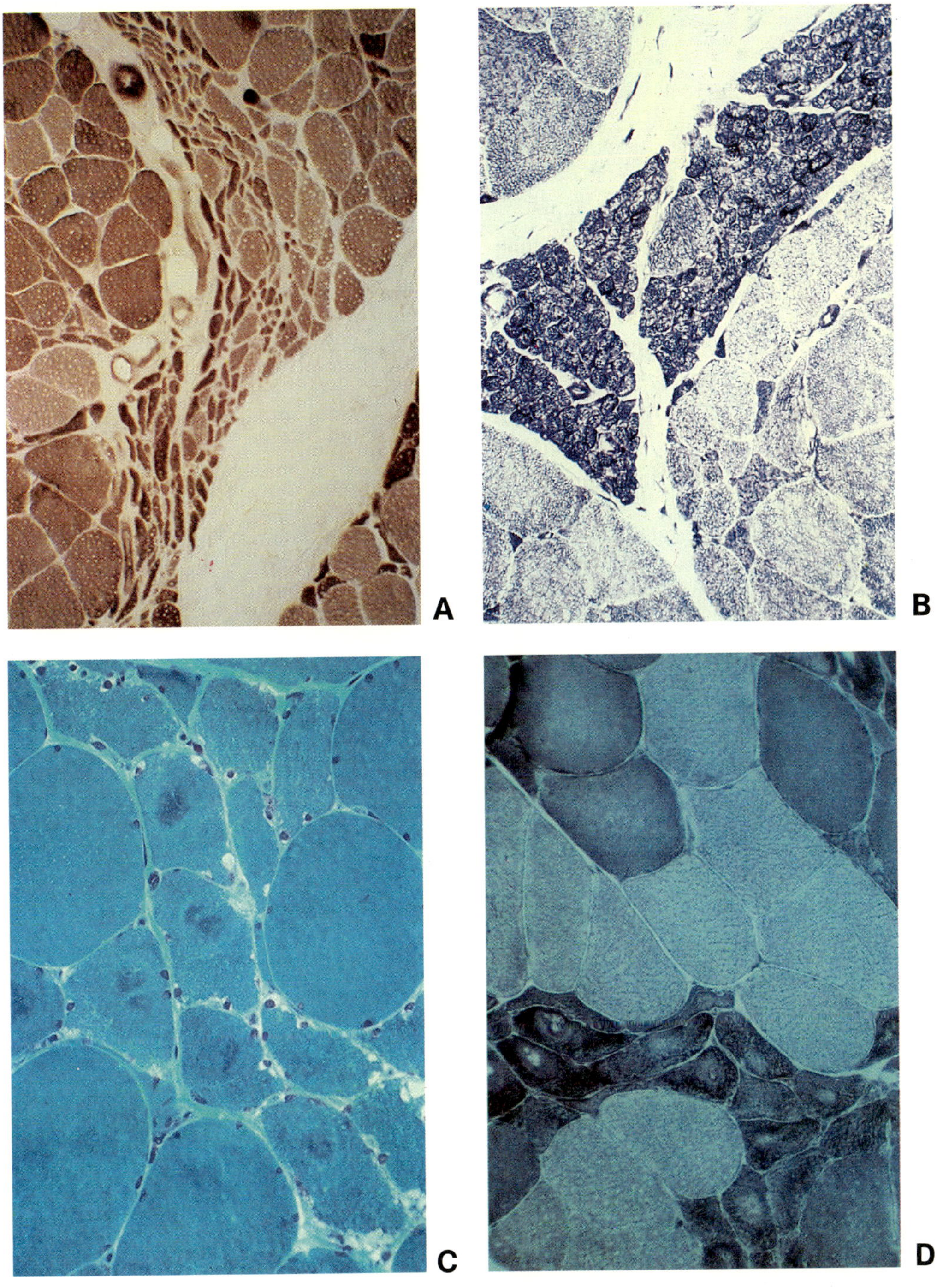

Plate 21 Cross sections of muscle showing denervation changes. **A:** Alkaline myofibrillar ATPase stain. **B,D:** NADH-TR stain. **C:** Modified Gomori trichrome stain. Atrophy of groups of fibers (group atrophy) is seen in denervation (**A,B**). Targets occur in fibers as the result of reinnervation, but for unknown reasons targets occur only in ATPase type I fibers (fibers dark with oxidative stains) and appear most readily with the NADH-TR stain. The three-zone target fiber has a central clear zone surrounded by a darker staining ring, and then normal fiber; hence, the appearance of a target (**D**). With the modified Gomori trichrome stain, targets are less readily seen but occasionally may be visible.

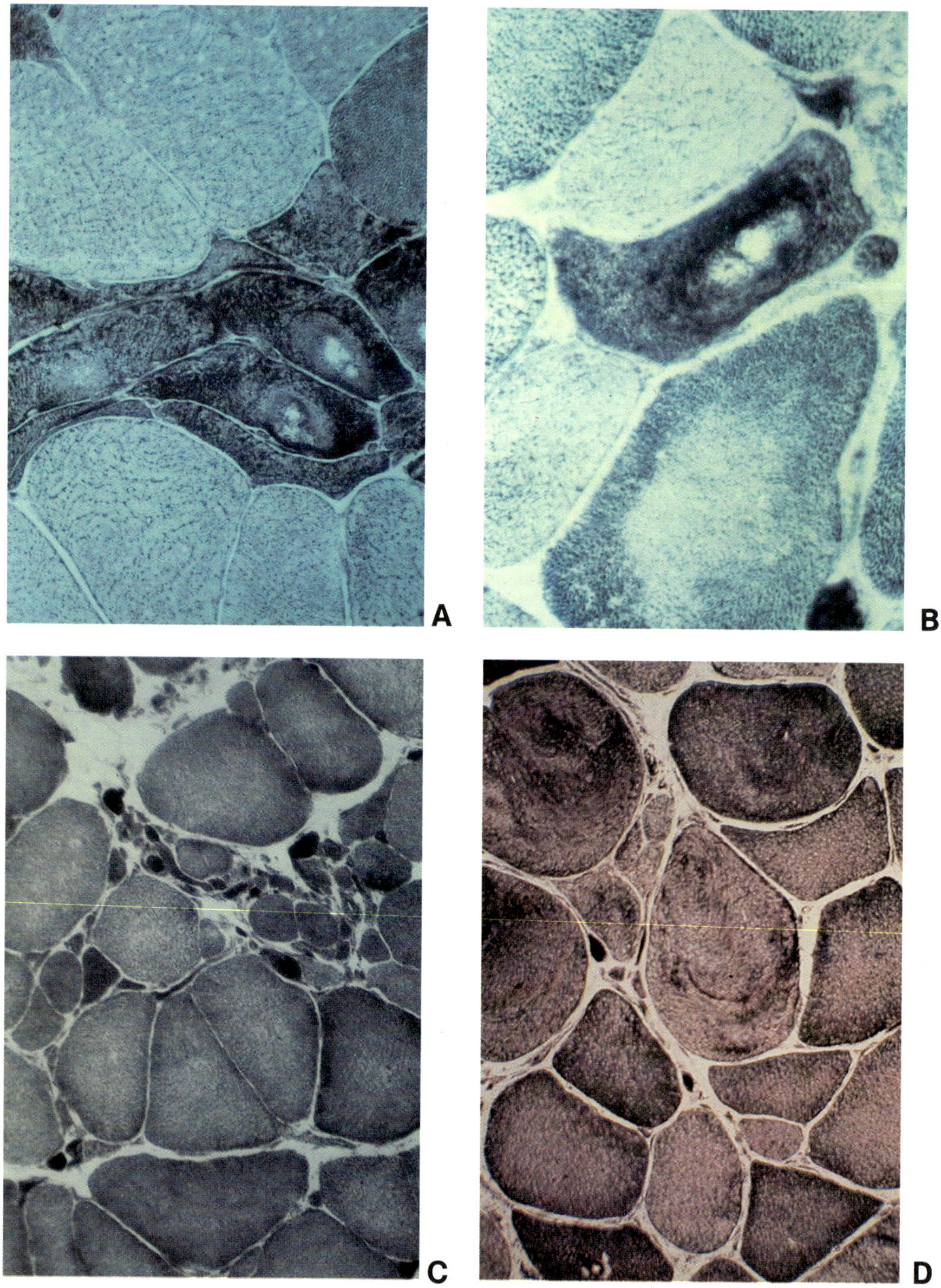

Plate 22 Cross sections of muscle showing denervation changes. **A,B,C,D:** NADH-TR stain. Target fibers have a single, usually central three-zone target (**A,B**). In denervation, there may also be disturbances in the staining of the intermyofibrillar network with a moth-eaten appearance (**C**) or whorls (**D**).

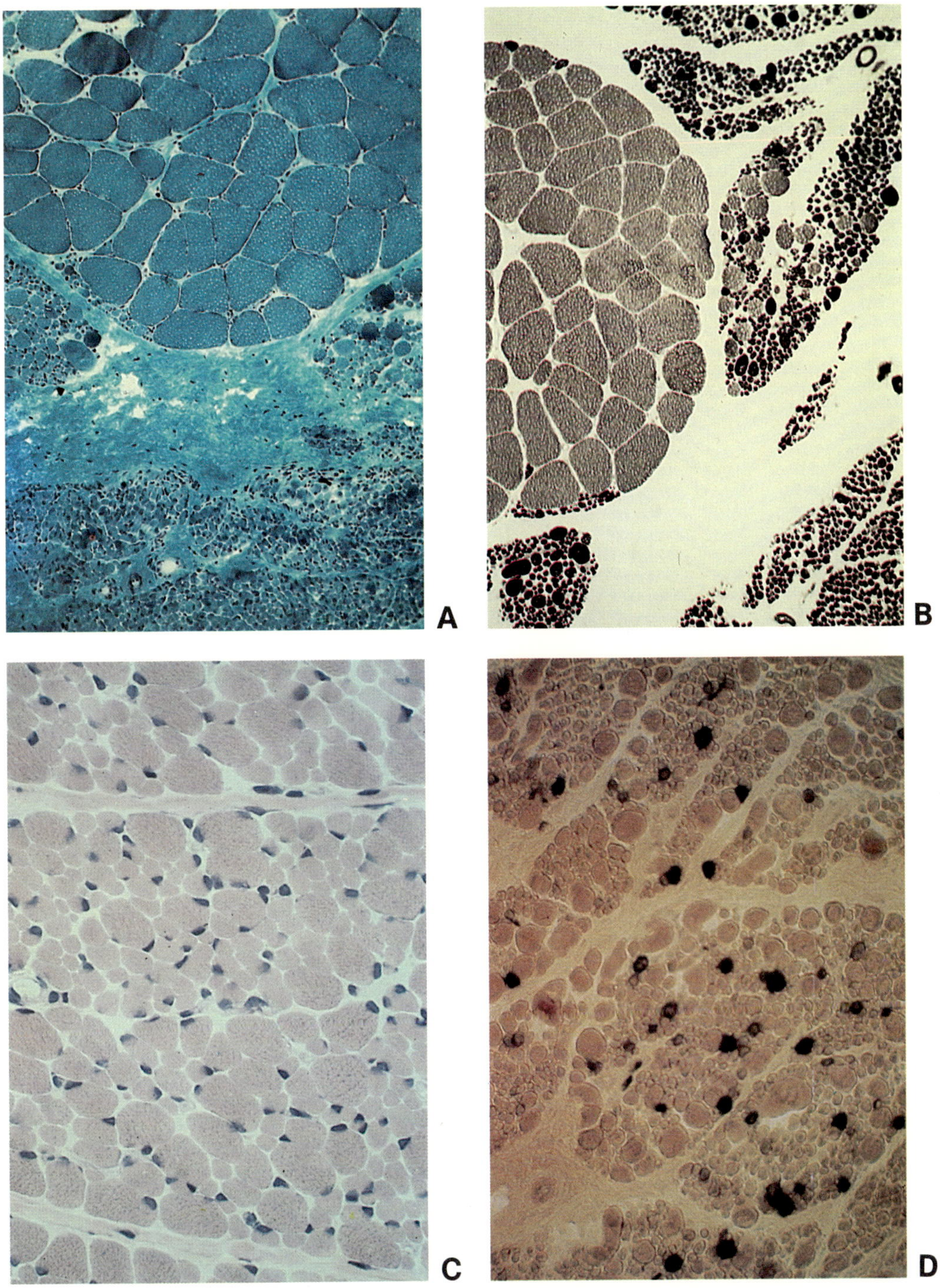

Plate 23 Cross sections of muscle from infants with infantile spinal muscular atrophy (Werdnig-Hoffmann disease). **A:** Modified Gomori trichrome stain. **B:** Alkaline myofibrillar ATPase stain. **C:** H&E stain. **D:** Alkaline phosphatase stain. Infantile spinal muscular atrophy is a special form of denervation beginning in utero in infants (inherited as an autosomal recessive disease). The findings in the biopsy are many very tiny rounded atrophic fibers (**A,B,C**), some of which stain positively with alkaline phosphatase stain (**D**), along with groups of very large (hypertrophied) fibers. The large hypertrophied fibers are all ATPase type I, while the small fibers are mainly ATPase type II.

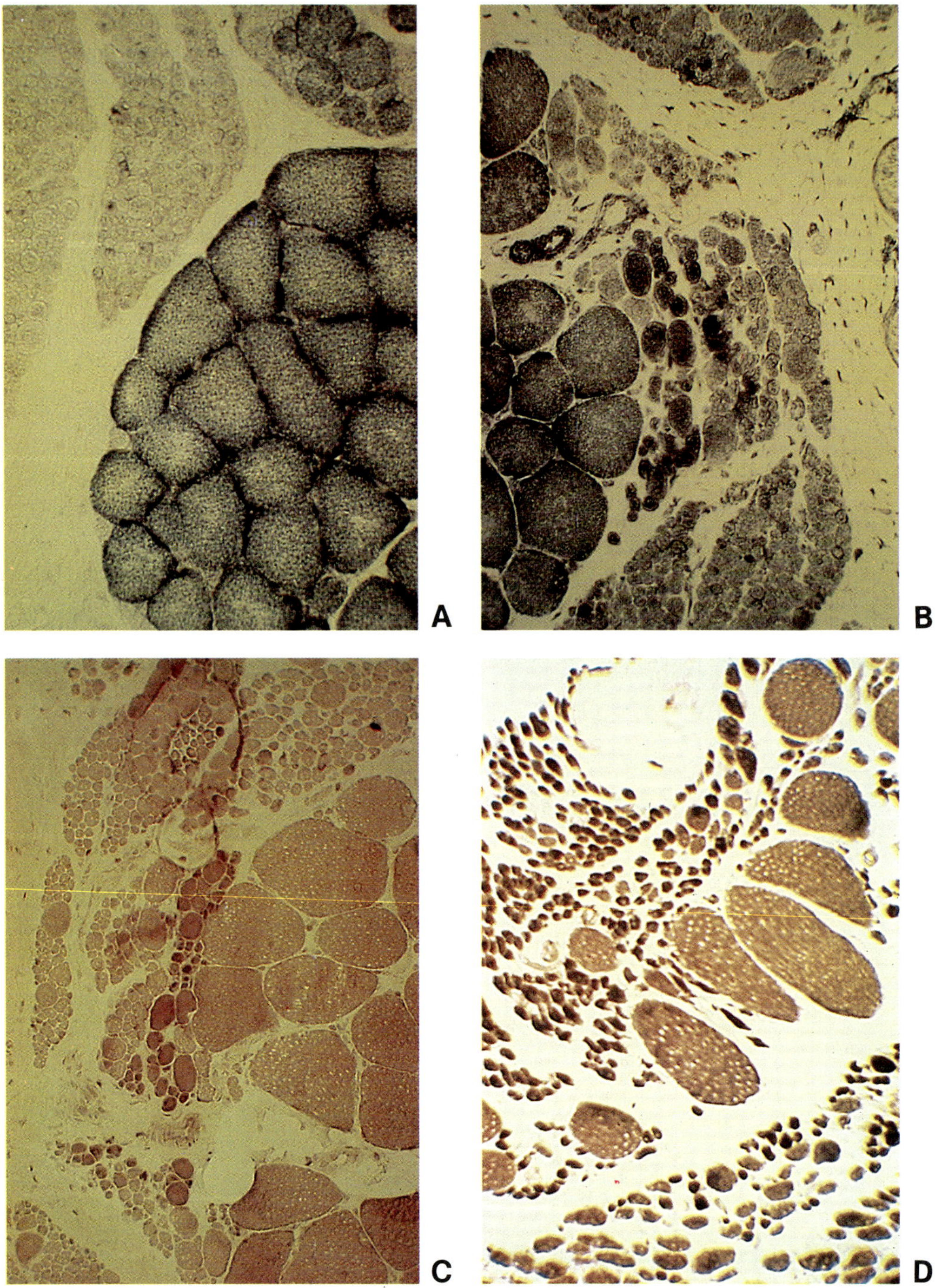

Plate 24 Cross sections of muscle from infants with infantile spinal muscular atrophy (Werdnig-Hoffmann disease). **A:** SDH stain. **B:** NADH-TR stain. **C:** Nonspecific esterase stain. **D:** Alkaline myofibrillar ATPase stain. This special form of denervation is characterized by large hypertrophied fibers with the histochemical features of ATPase type I fibers and very tiny rounded fibers which are mainly ATPase type II fibers.

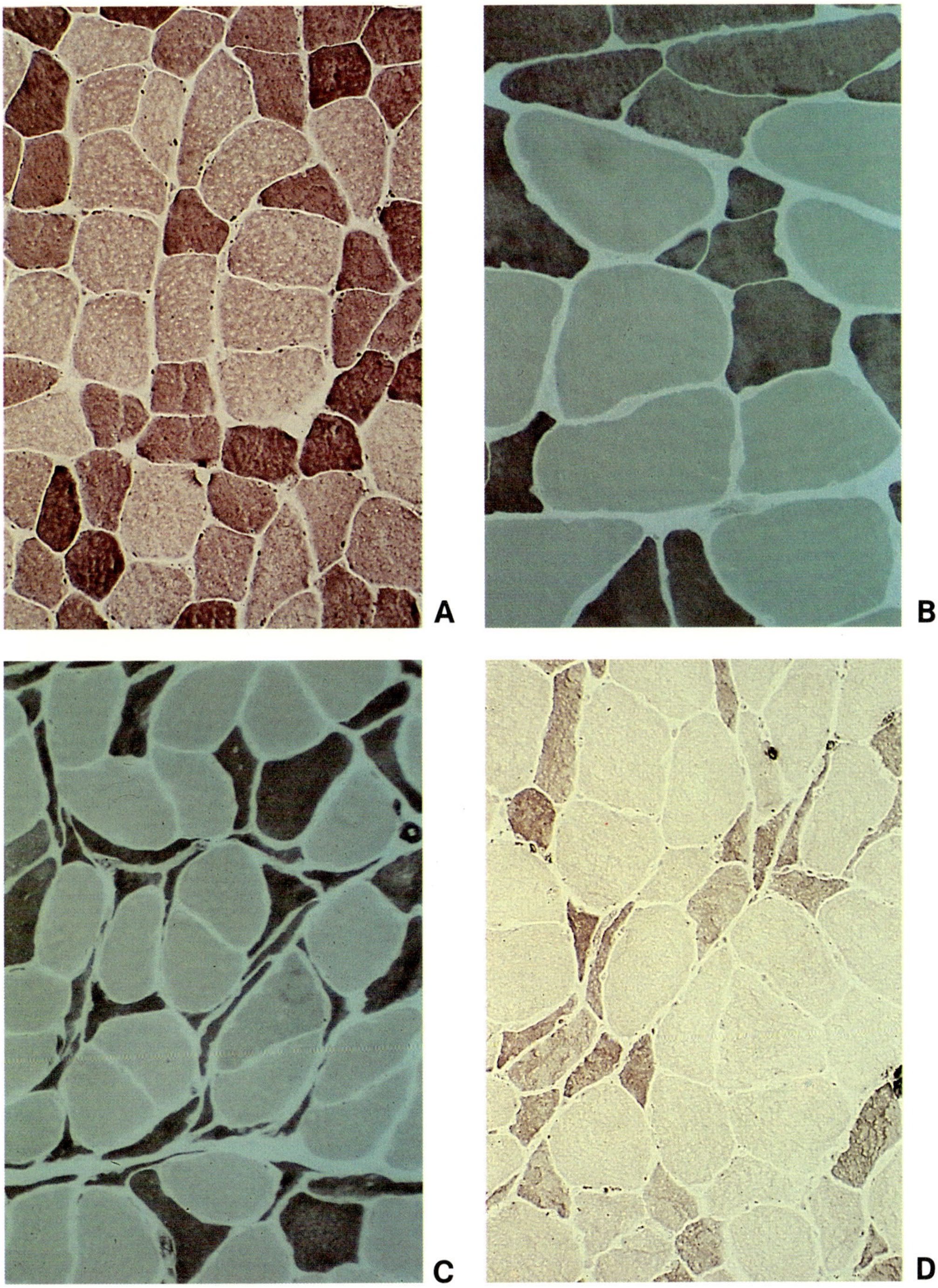

Plate 25 Cross sections of muscle specimens with type II muscle fiber atrophy. **A,B,C,D:** Alkaline myofibrillar ATPase stain. This is one of the most common causes of muscle weakness in hospitalized or chronically ill patients. This is clinically sometimes called (based upon the particular illness of the patient) hypothyroid myopathy, hypoparathyroid myopathy, hyperthyroid myopathy, hyperparathyroid myopathy, steroid myopathy, cachexic myopathy, carcinomatous myopathy, or disuse atrophy. Type II fiber atrophy is sometimes histologically confused with denervation because the atrophic fibers may be angular. A key factor in distinguishing between them is that the small fibers are ATPase type II only in type II fiber atrophy, whereas in denervation there are small fibers of both fiber types.

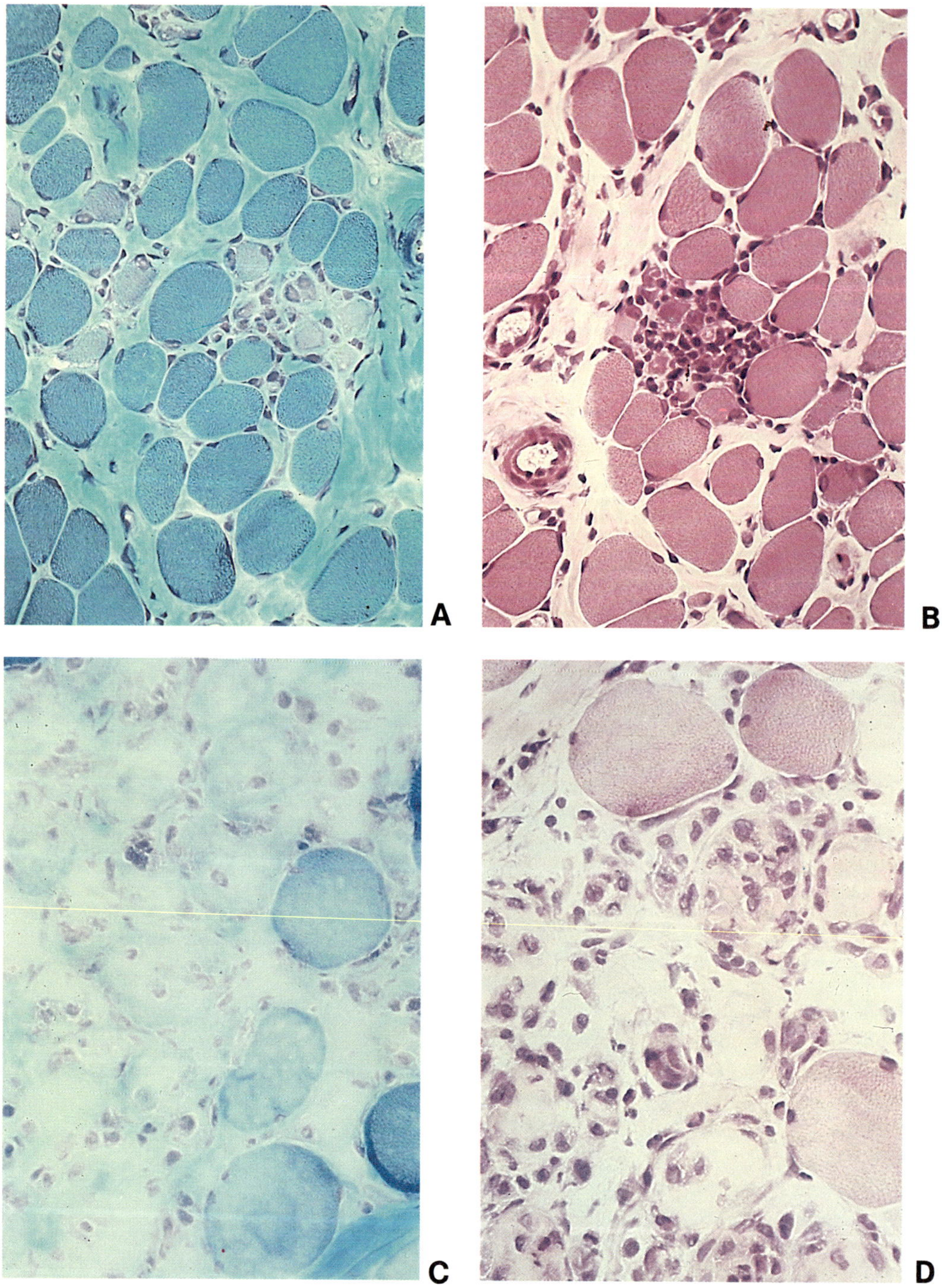

Plate 26 Cross sections of muscle from patients with Duchenne muscular dystrophy. **A,C:** Modified Gomori trichrome stain. **B,D:** H&E stain. Focal collections of inflammatory cells, fiber size variation, degenerating fibers and marked increase in endomysial connective tissues are characteristic.

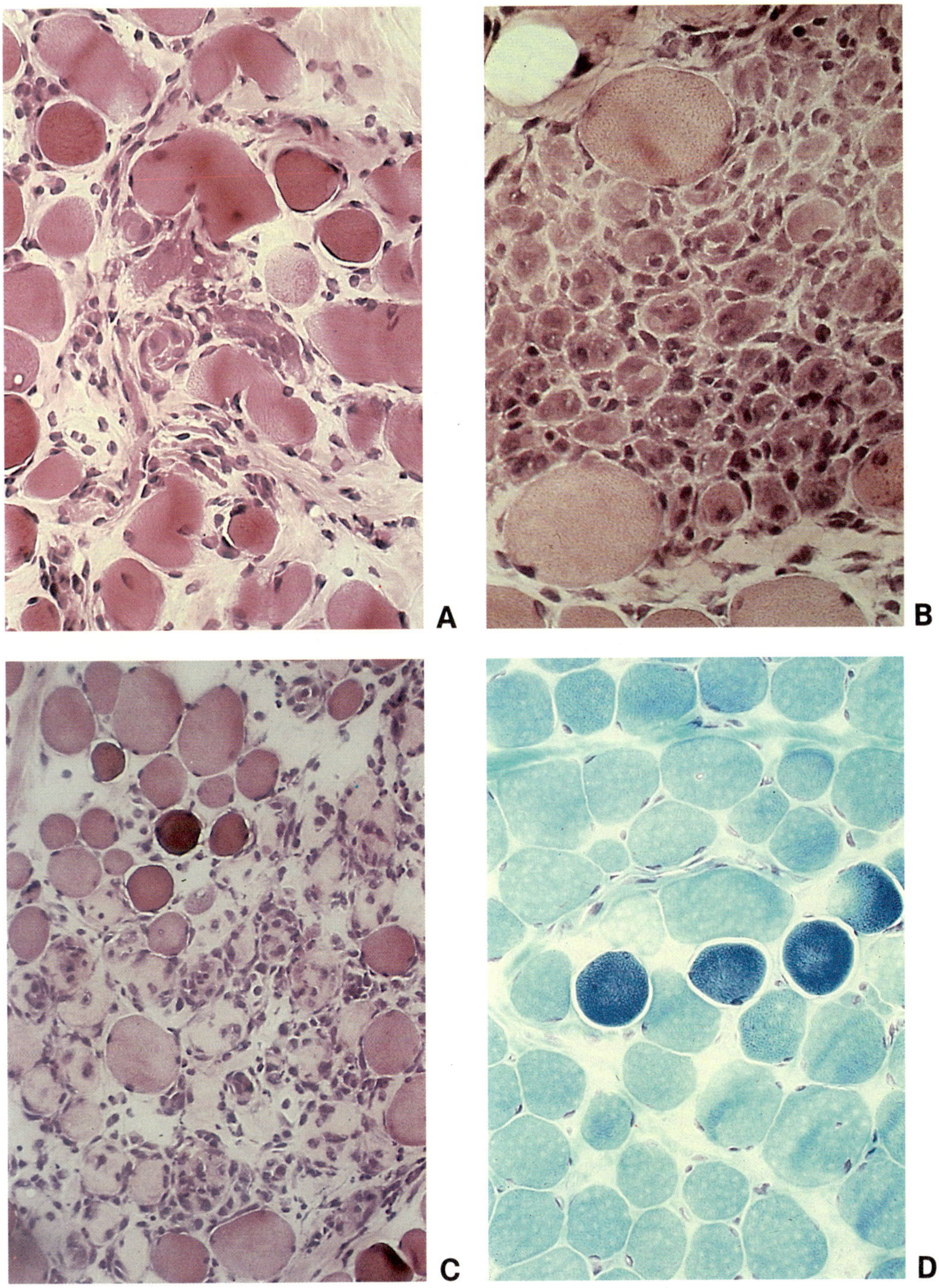

Plate 27 Cross sections of muscle from patients with Duchenne muscular dystrophy. **A,B,C:** H&E stain. **D:** Modified Gomori trichrome stain. There are variations in fiber size (**A**), basophilic regenerating fibers (**B,C**), and large, excessively dark, hypercontracted fibers (**C,D**).

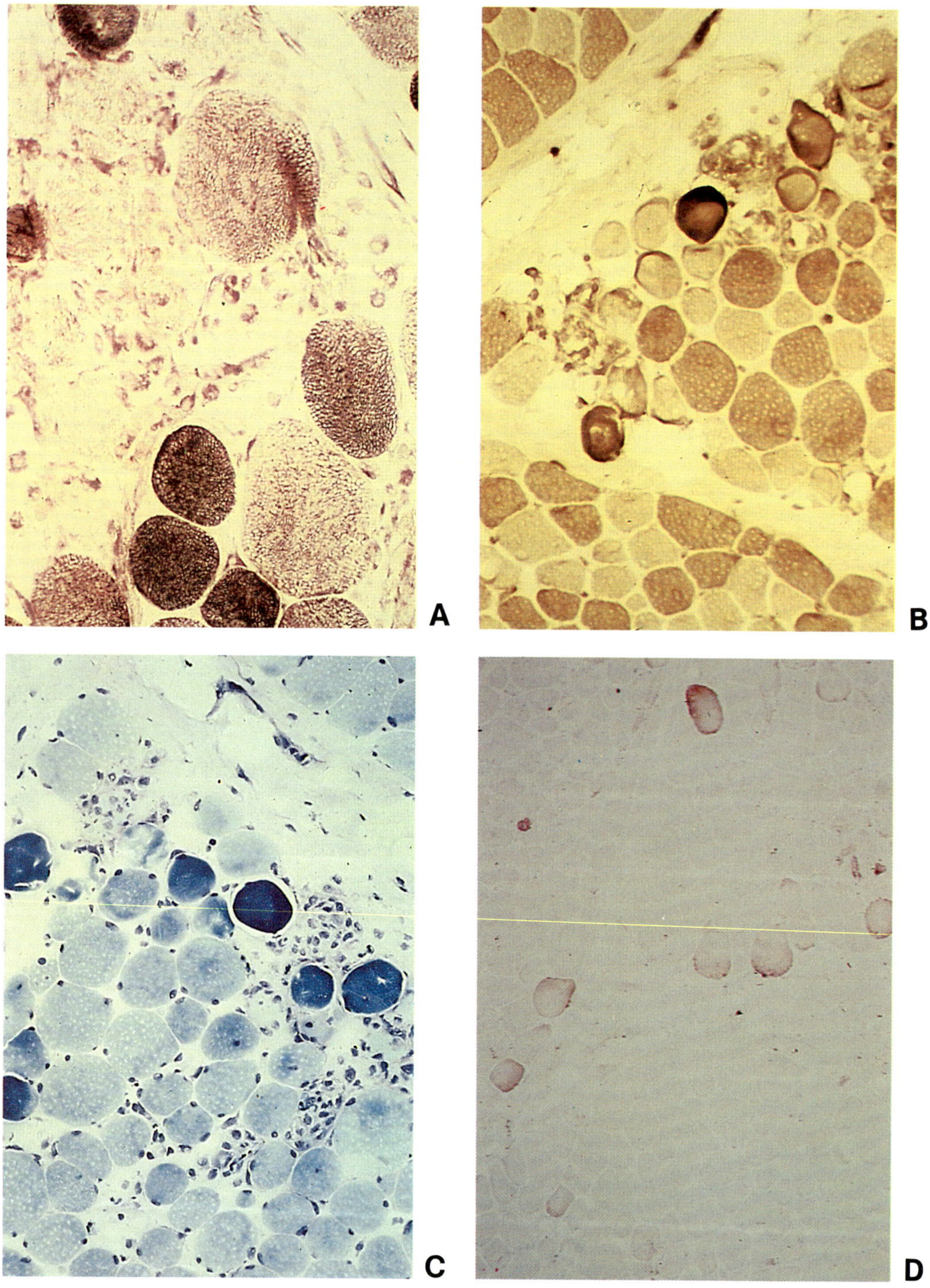

Plate 28 Cross sections of muscle specimens from patients with Duchenne muscular dystrophy. **A:** NADH-TR stain. **B:** Alkaline myofibrillar ATPase stain. **C:** Toluidine blue stain. **D:** Alizarin red S stain. The large dark hypercontracted fibers are evident with multiple stains (**A,B,C**) and stain positively for excess calcium with the alizarin red S stain (**D**).

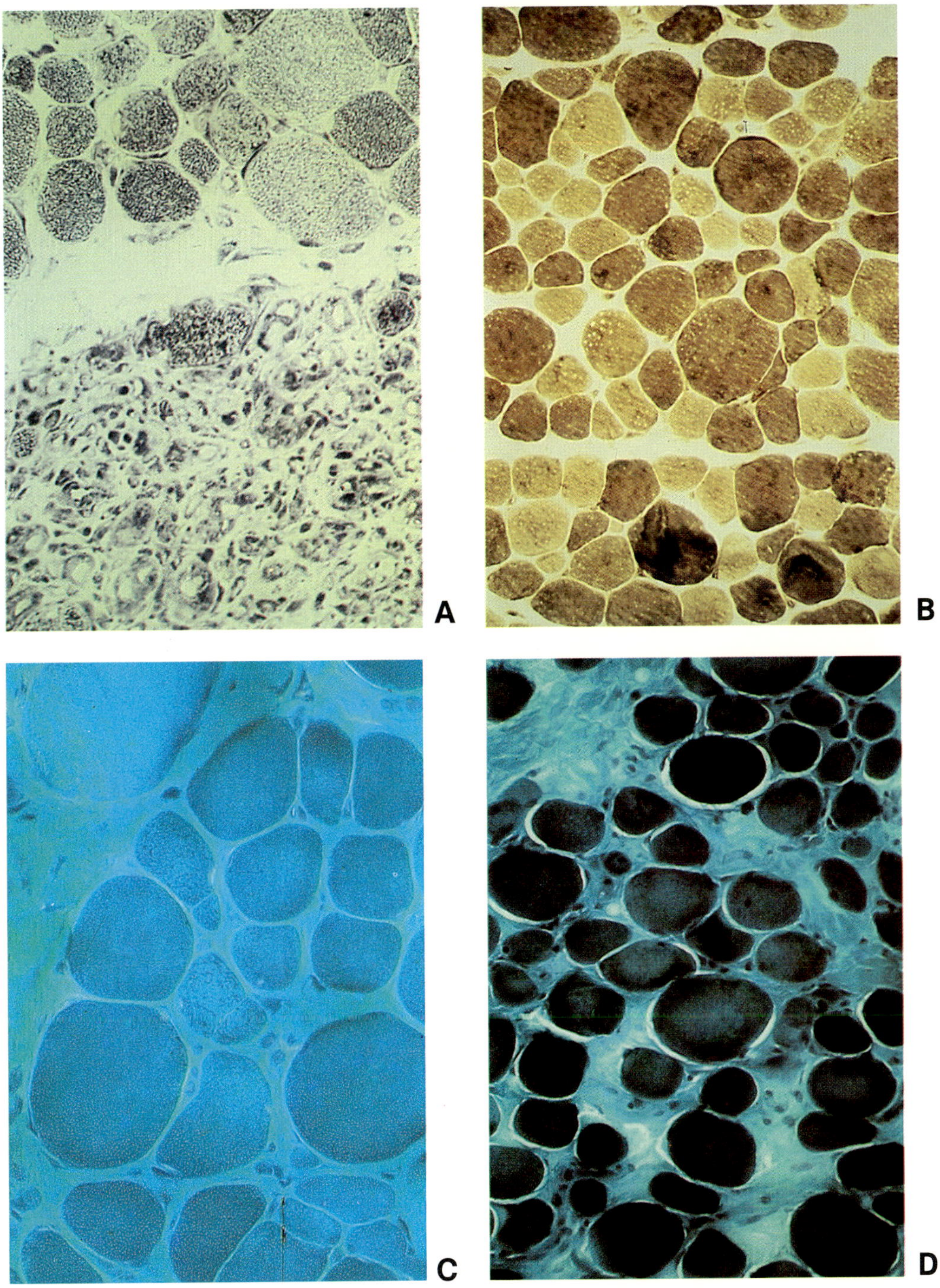

Plate 29 Cross sections of muscle specimens from patients with Duchenne muscular dystrophy. **A:** NADH-TR stain. **B:** Alkaline myofibrillar ATPase stain. **C,D:** Modified Gomori trichrome stain. Both type I and type II muscle fibers are equally affected. Note the marked increase in endomysial connective tissue (unstained by the NADH-TR and ATPase stains and green with the modified Gomori trichrome stain) separating muscle fibers.

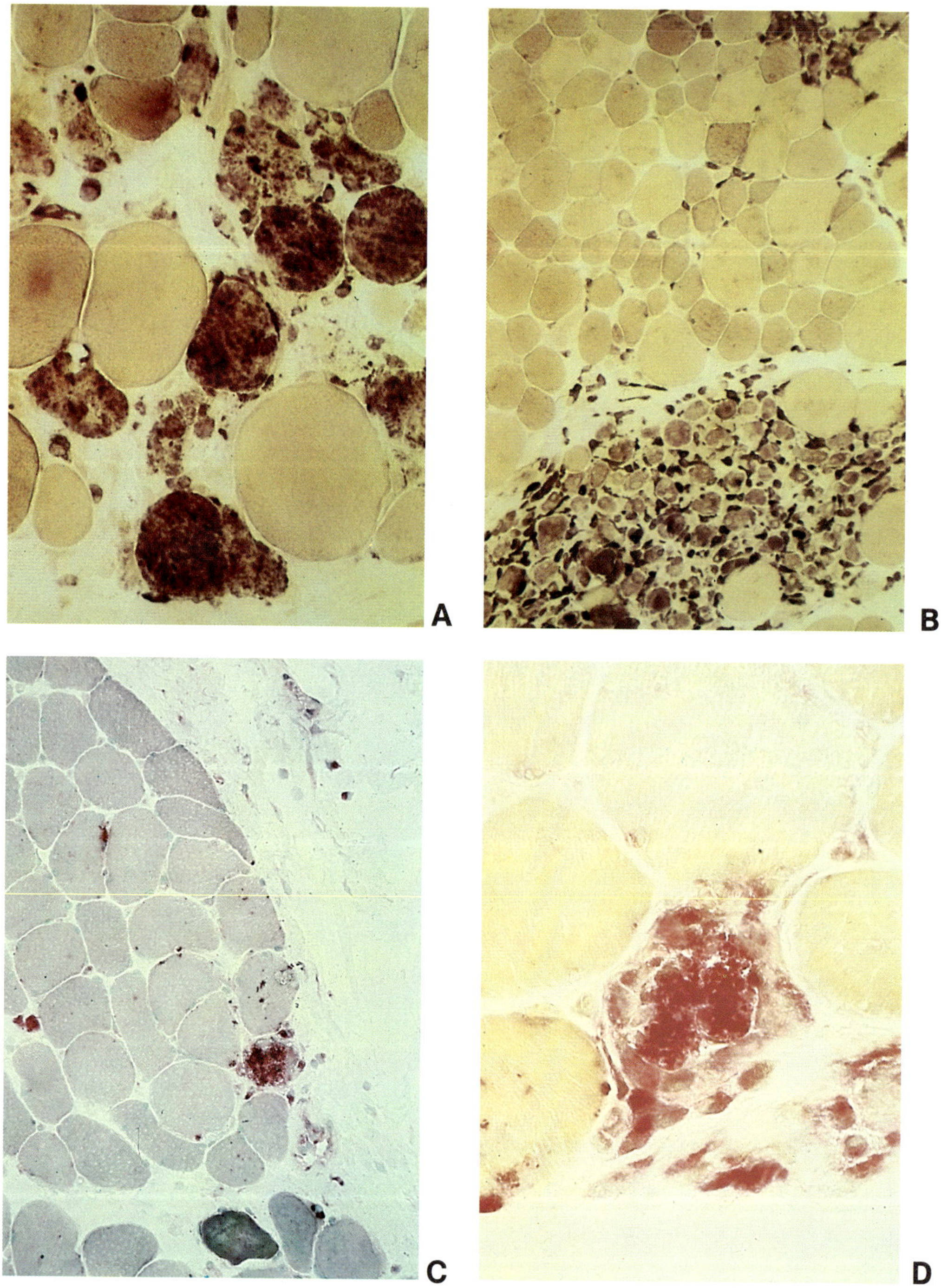

Plate 30 Cross sections of muscle specimens from patients with Duchenne muscular dystrophy. **A,B:** Nonspecific esterase. **C,D:** Acid phosphatase. Macrophages infiltrating between fibers and digesting damaged fibers stain dark with nonspecific esterase stain and are red with acid phosphatase stain.

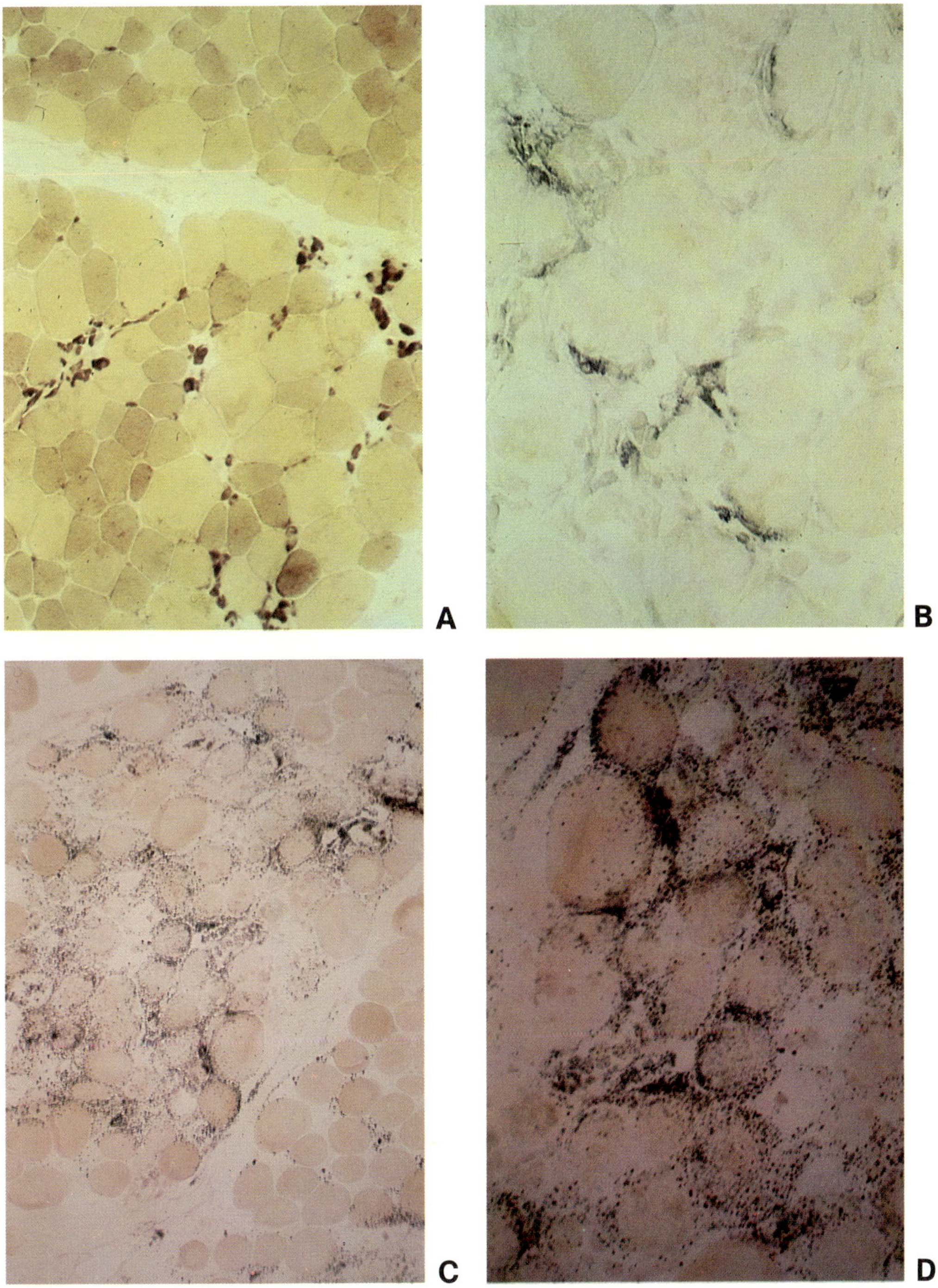

Plate 31 Cross sections of muscle specimens from patients with Duchenne muscular dystrophy. **A:** Nonspecific esterase stain. **B,C,D:** Alkaline phosphatase stain. Macrophages in the endomysial spaces stain dark with the nonspecific esterase stain (**A**). The excessive fibroblast activity producing the characteristic marked proliferation of connective tissue in Duchenne muscular dystrophy is associated with dark alkaline phosphatase staining of the connective tissue (**B,C,D**).

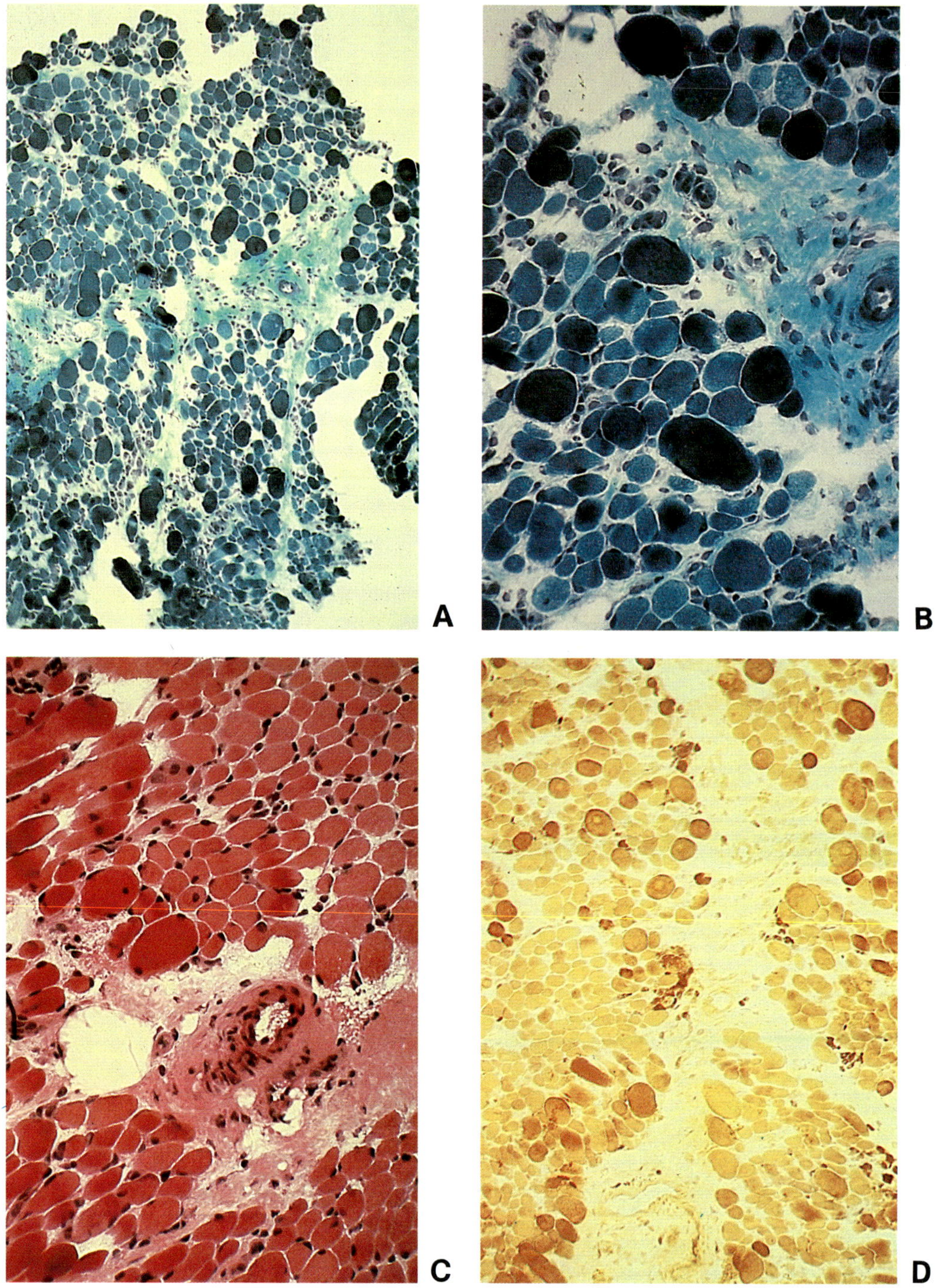

Plate 32 Cross sections of muscle specimens from 4-month-old infant with Duchenne muscular dystrophy. **A,B:** Modified Gomori trichrome stain. **C:** H&E stain. **D:** Nonspecific esterase stain. This disorder can be identified in genetically at-risk infants who appear clinically normal but have elevated serum creatine phosphokinase levels. Despite the normally small diameter (10 to 15 μm) of infant muscle fibers, there is visible variability in fiber size, occasional large, dark hypercontracted fibers, and proliferation of connective tissue.

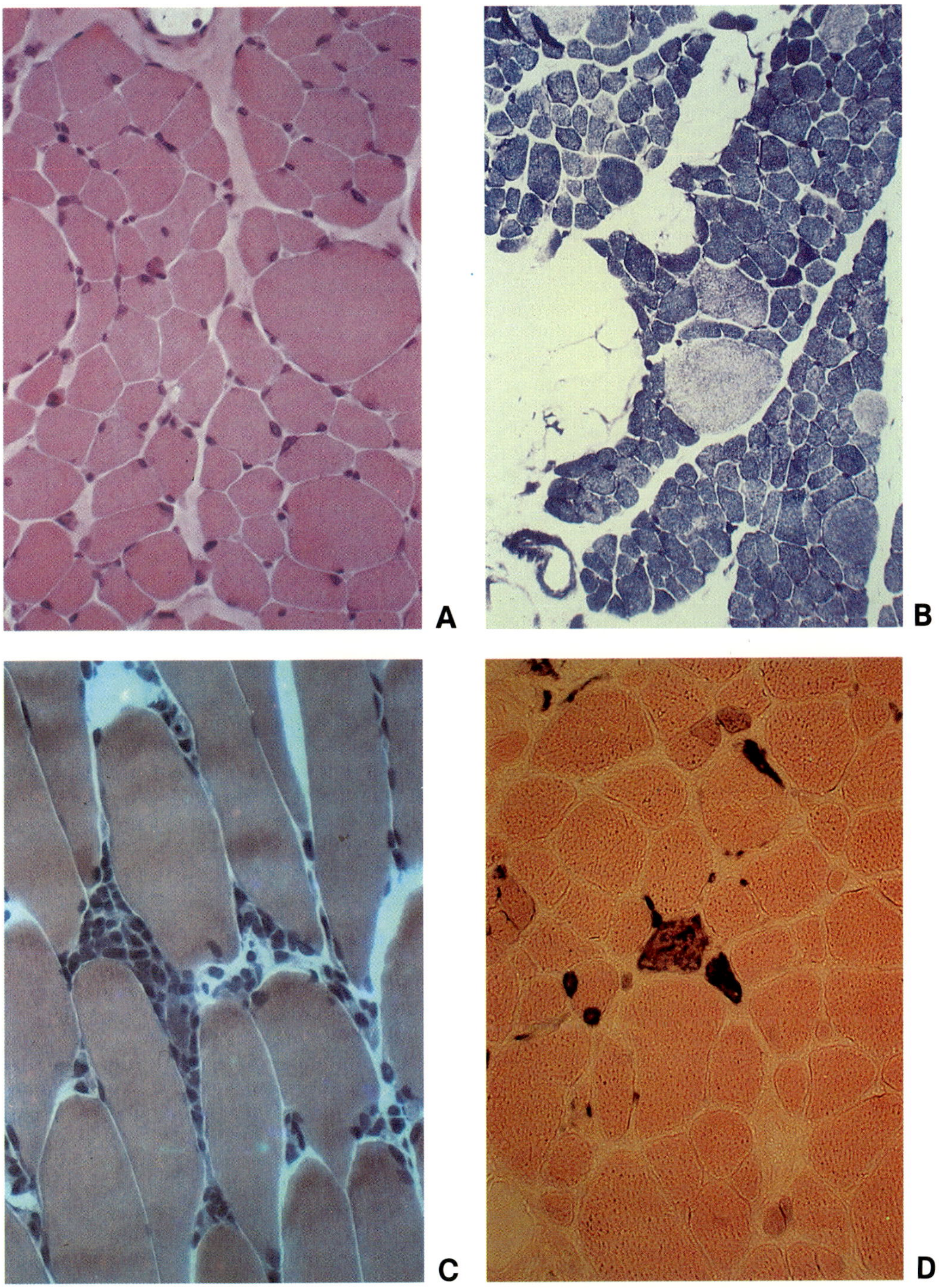

Plate 33 Cross sections of muscle specimens from patients with fascioscapulohumeral dystrophy. **A,C:** H&E stain. **B:** NADH-TR stain. **D:** Alkaline phosphatase stain. This disease is characterized by both myopathic and neuropathic features. The myopathic features include hypertrophied and small, rounded fibers (**A**), focal inflammatory cell infiltrates (**C**) and degenerating fibers which stain dark with the alkaline phosphatase stain (**D**). The neuropathic features include angular fibers which are dark with NADH-TR (**B**) and nonspecific esterase.

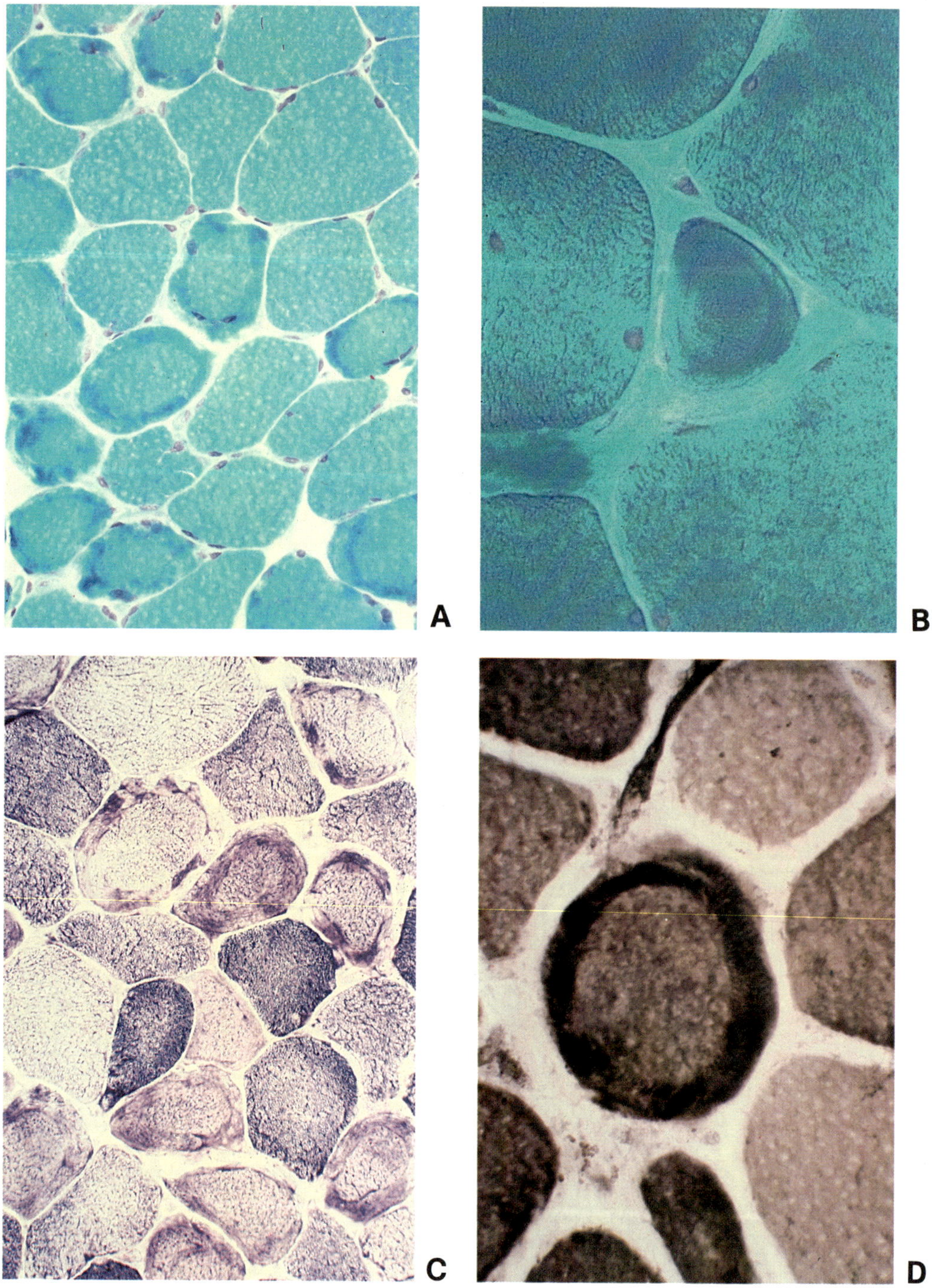

Plate 34 Muscle biopsy specimens showing ring fibers. **A,B:** Modified Gomori trichrome stain. **C:** NADH-TR stain. **D:** Alkaline myofibrillar ATPase stain. The normally oriented myofibrils are distorted by myofibrils oriented at right angles to the main body of the muscle fiber. Ring fibers are most common in myotonic dystrophy but may be seen in many myopathies.

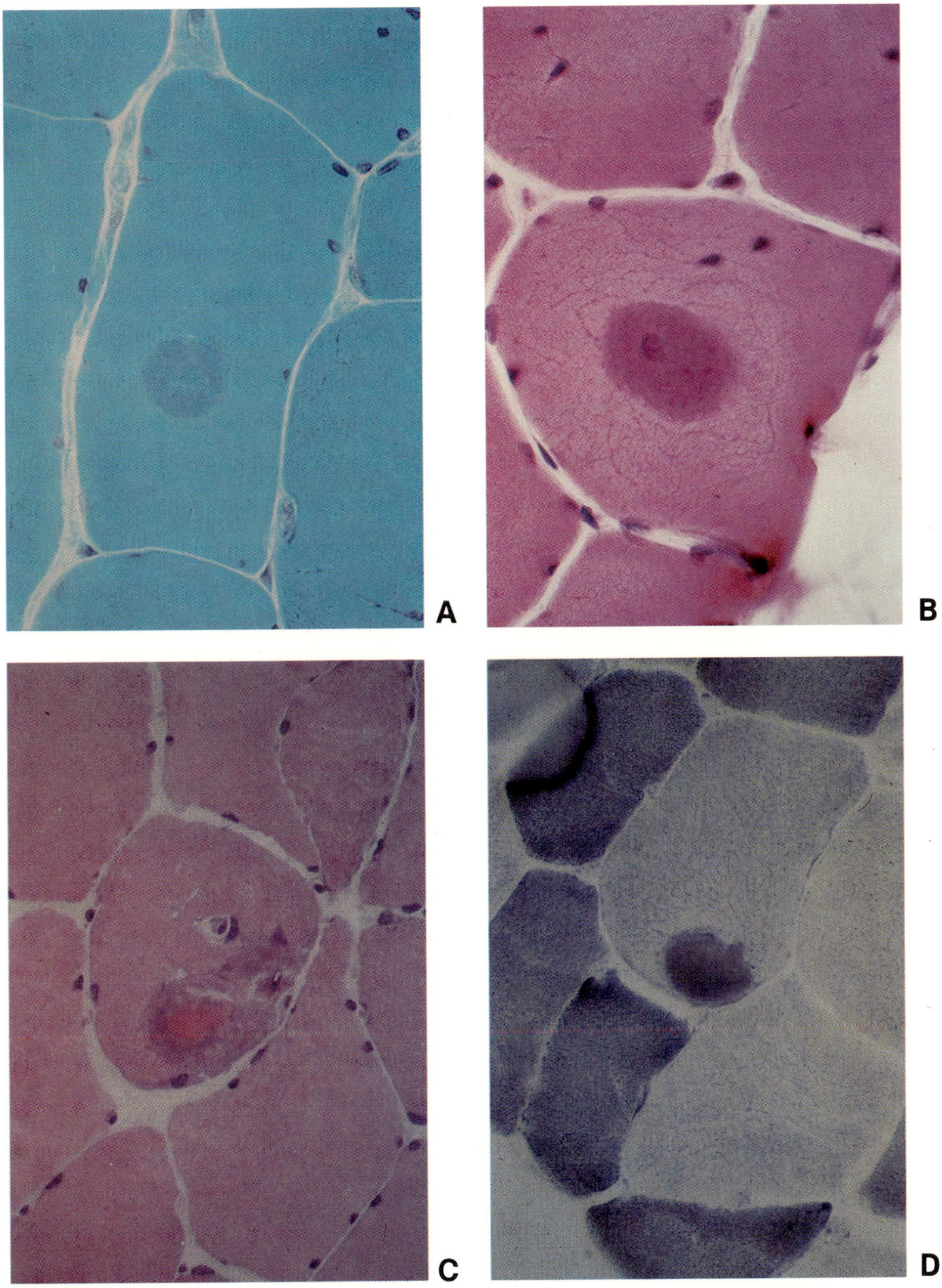

Plate 35 Muscle biopsy specimens showing cytoplasmic bodies. **A:** Modified Gomori trichrome stain. **B,C:** H&E stain. **D:** NADH-TR stain. The exact composition of cytoplasmic bodies is unknown.

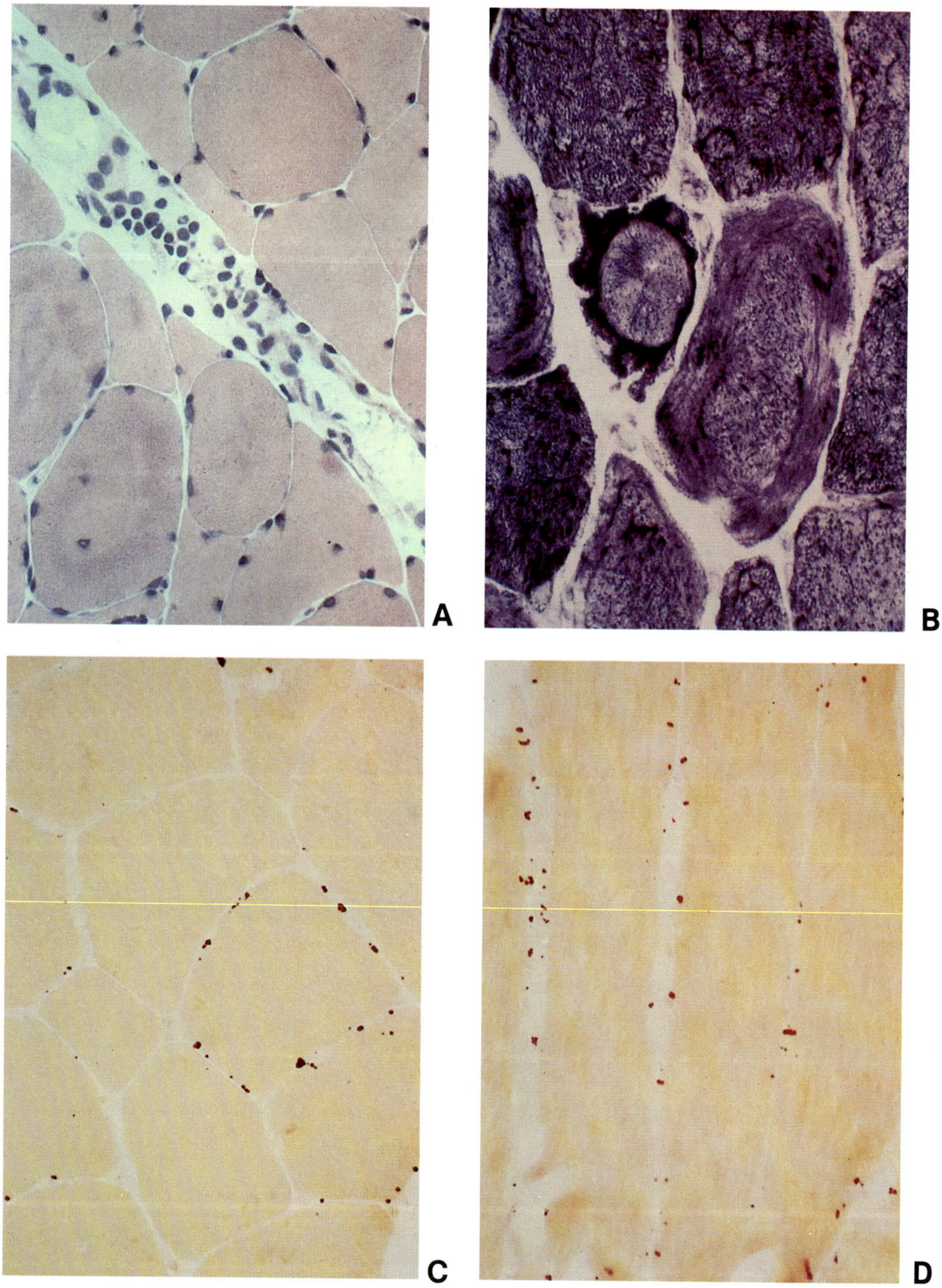

Plate 36 Muscle biopsy specimens showing nonspecific myopathic changes. **A:** H&E stain. **B:** NADH-TR stain. **C,D:** Nonspecific esterase stain. Disturbed fiber architecture, including rings and ragged red fibers (**A,B**) are occasionally seen as nonspecific features of myopathy. Excessive accumulation of lipofuscin in the subsarcolemmal areas (stained as dark dots by nonspecific esterase stain) is occasionally seen in patients with vague clinical symptomatology of weakness.

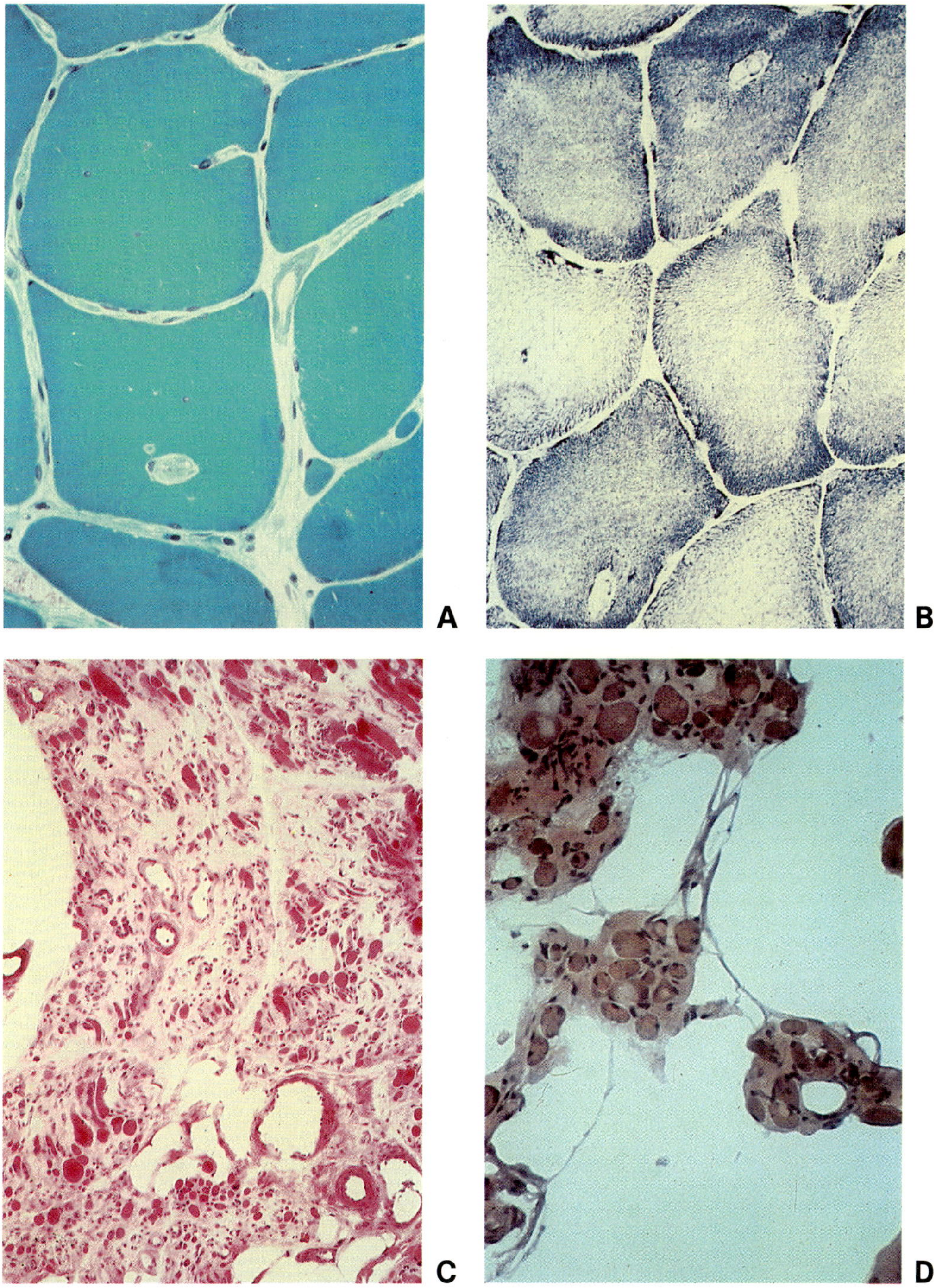

Plate 37 Muscle specimens showing myopathic changes. **A:** Modified Gomori trichrome stain. **B:** NADH-TR stain. **C,D:** H&E stain. Internalized capillaries (**A,B**) are a nonspecific myopathic feature. End-stage muscle (**C,D**) from whatever process (either myopathic or neuropathic disease) demonstrates few muscle fibers with connective tissue and fat replacement. The histological picture of end-stage muscle can be seen in the presence of clinical strength of complete extremity movement against gravity (Grade 3 on the MRC scale).

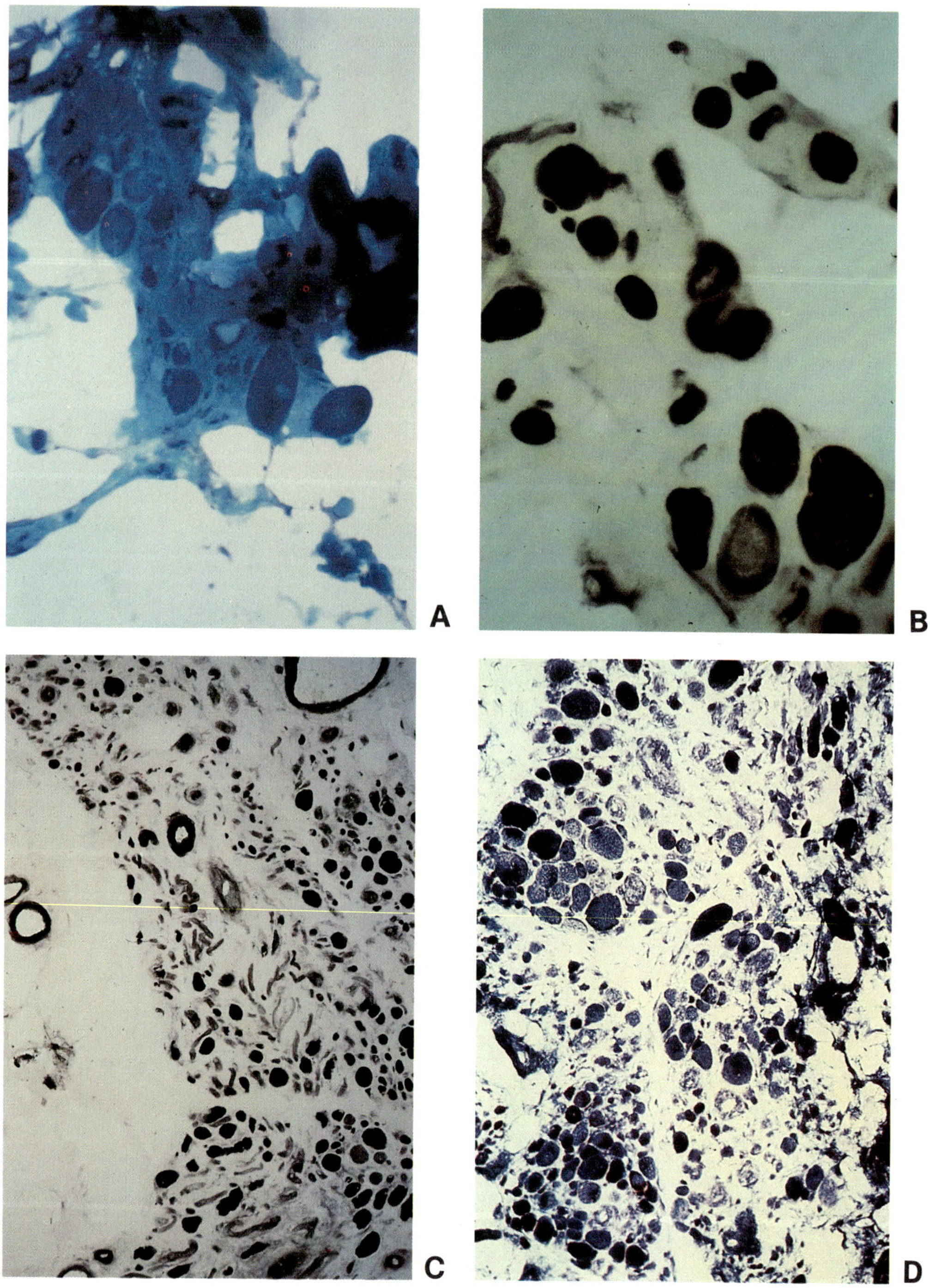

Plate 38 Muscle specimens showing histologically end-stage muscle. **A:** Modified Gomori trichrome. **B,C:** Alkaline myofibrillar ATPase. **D:** NADH-TR. The few fibers that remain in end-stage muscle specimens display poor histochemical differentiation.

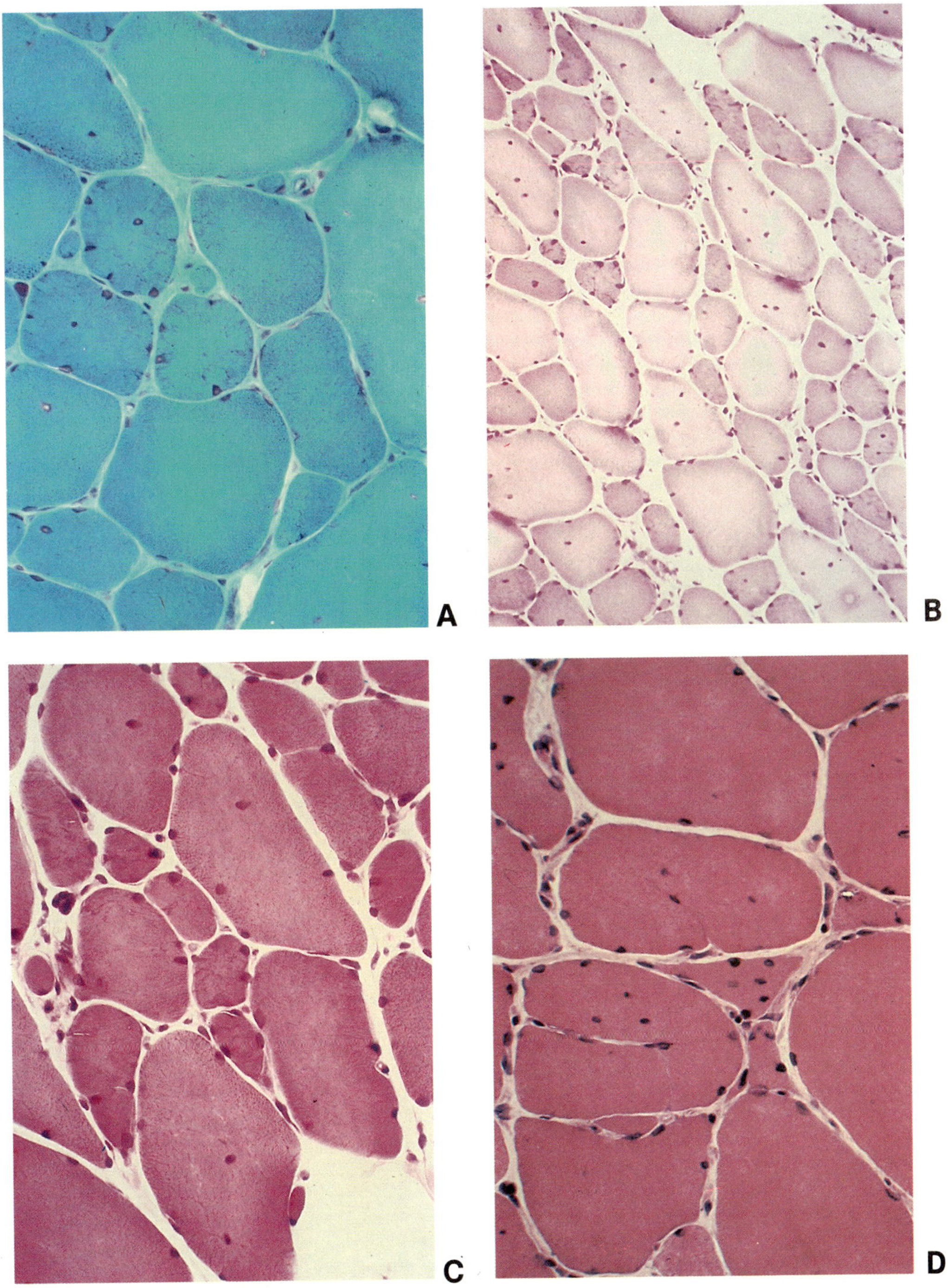

Plate 39 Muscle specimens from patients with limb-girdle dystrophy. **A:** Modified Gomori trichrome stain. **B,C,D:** H&E stain. There is variation in fiber size, many internal nuclei, and an increase in endomysial connective tissue.

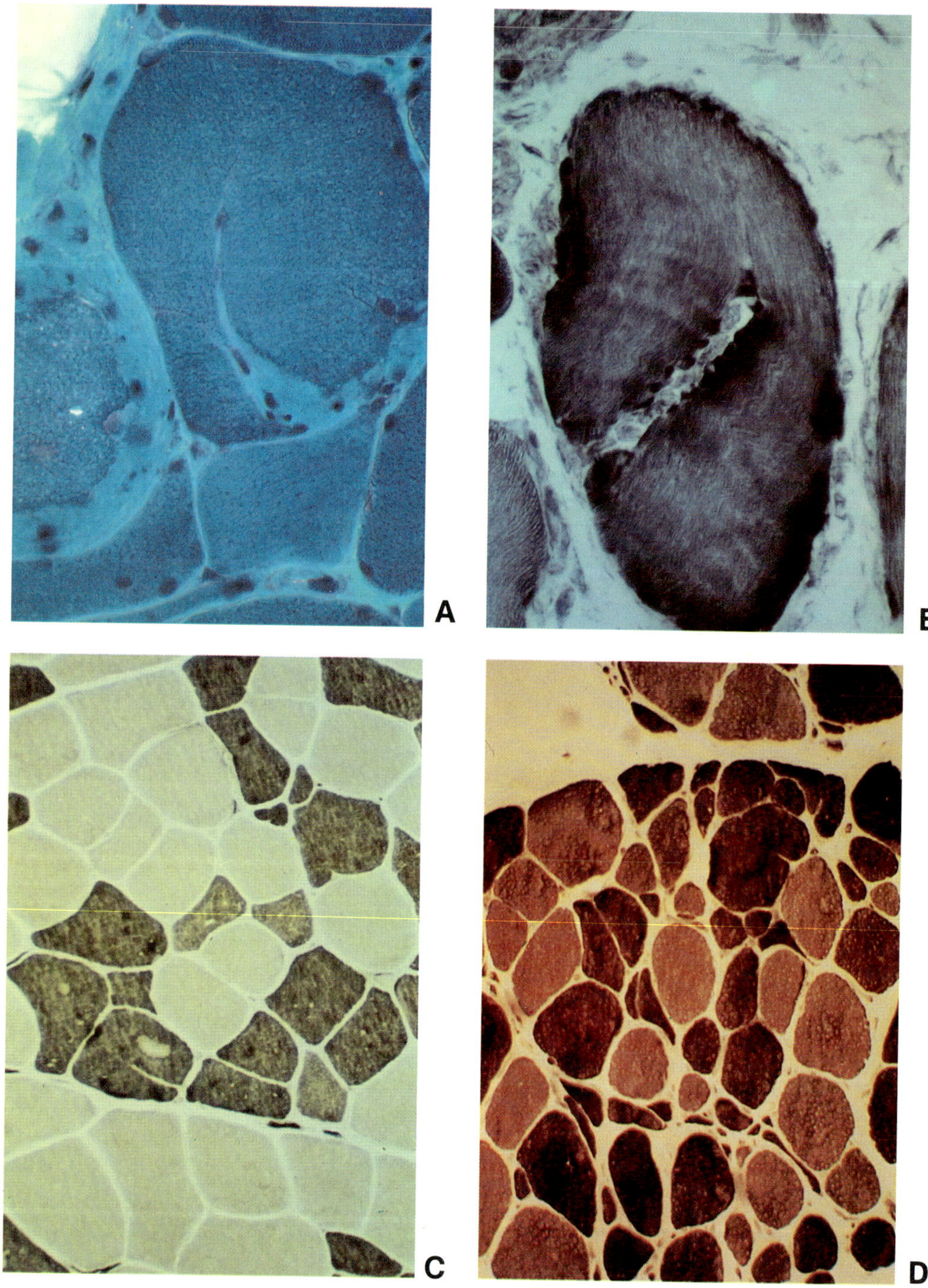

Plate 40 Muscle specimens from patients with limb-girdle dystrophy. **A:** Modified Gomori trichrome stain. **B:** NADH-TR stain. **C,D:** Alkaline myofibrillar ATPase stain. Splitting of large fibers is evident (**A,B,C**). Variability in fiber size is characteristic (**A,C,D**).

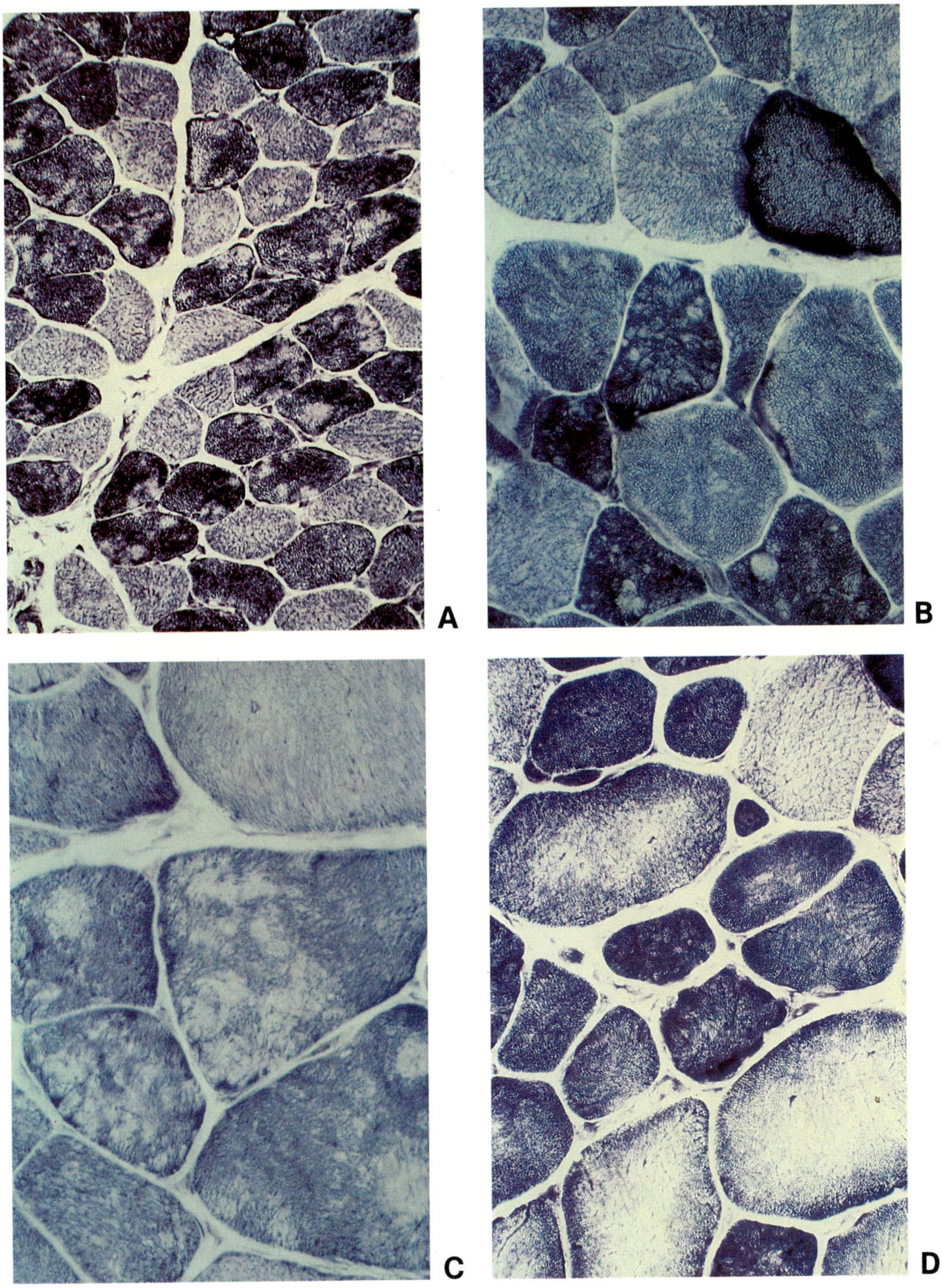

Plate 41 Muscle specimens from patients with limb-girdle dystropy **A,B,C,D.** NADH-TR stain. Many moth-eaten fibers (**A,B,C**) and fibers with clear centers (**D**) are characteristic of this disorder.

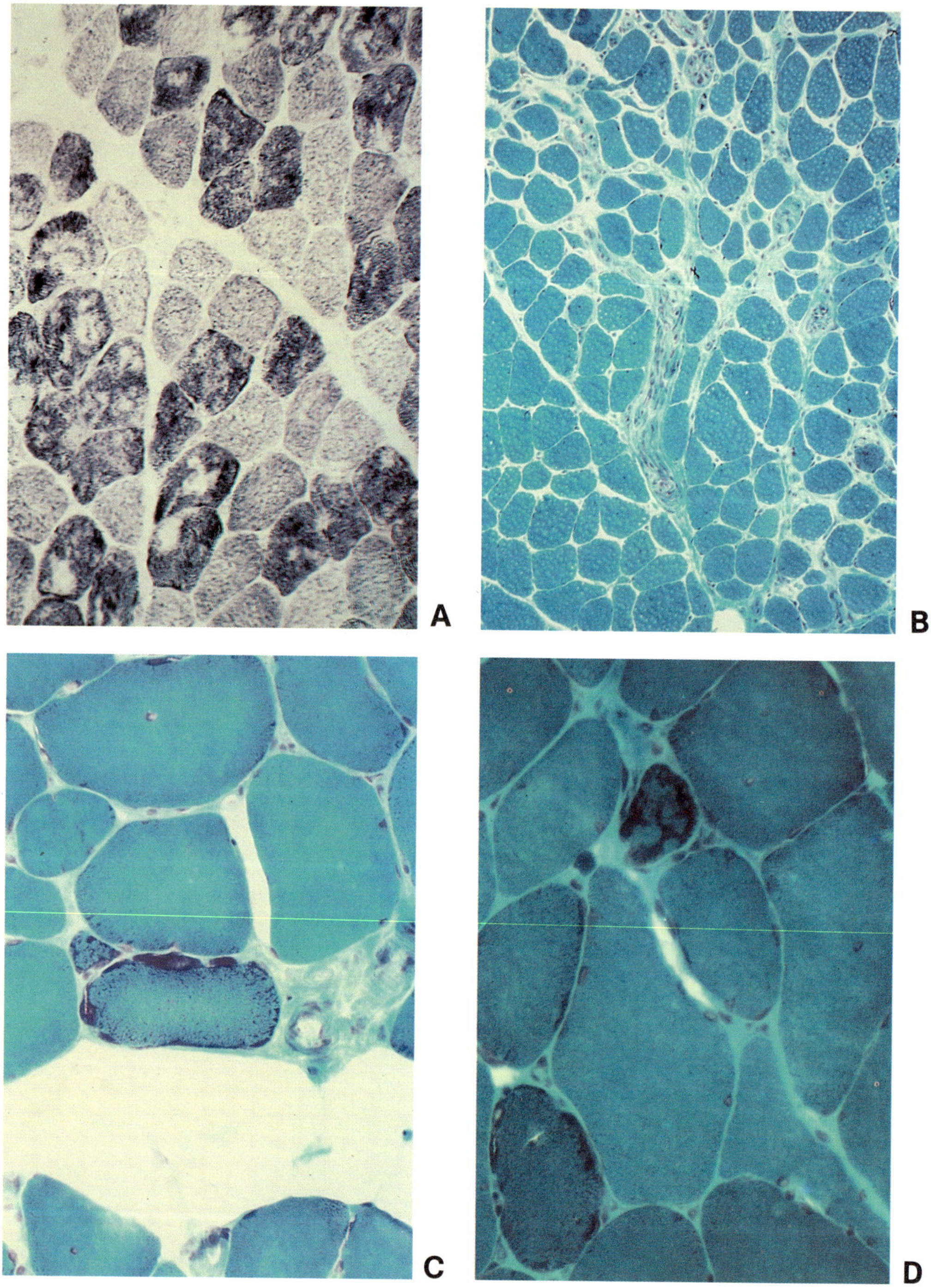

Plate 42 Muscle specimens from patients with limb-girdle dystrophy. **A:** SDH stain. **B,C,D:** Modified Gomori trichrome stain. "Moth-eaten" fibers are as evident with SDH (**A**) as with NADH-TR stains. Focal increases in connective tissue (**B**) and occasional ragged red fibers are also common (**C,D**).

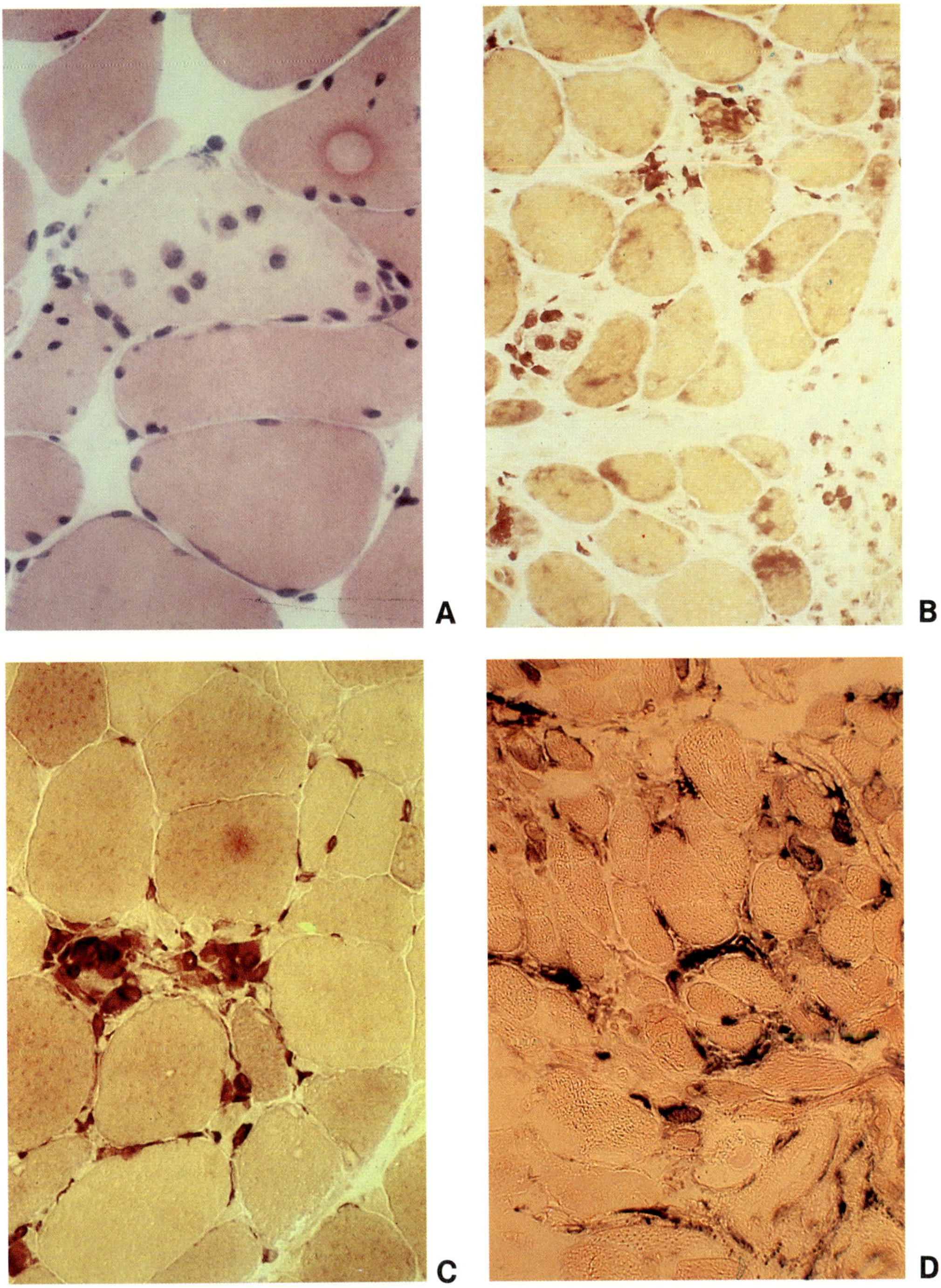

Plate 43 Muscle specimens from patients with limb-girdle dystrophy. **A:** H&E stain. **B,C:** Nonspecific esterase stain. **D:** Alkaline phosphatase stain. Focal degenerating fibers (**A**), focal inflammatory cell infiltrates (**B,C**), and alkaline phosphatase staining of connective tissue are also found with this disorder.

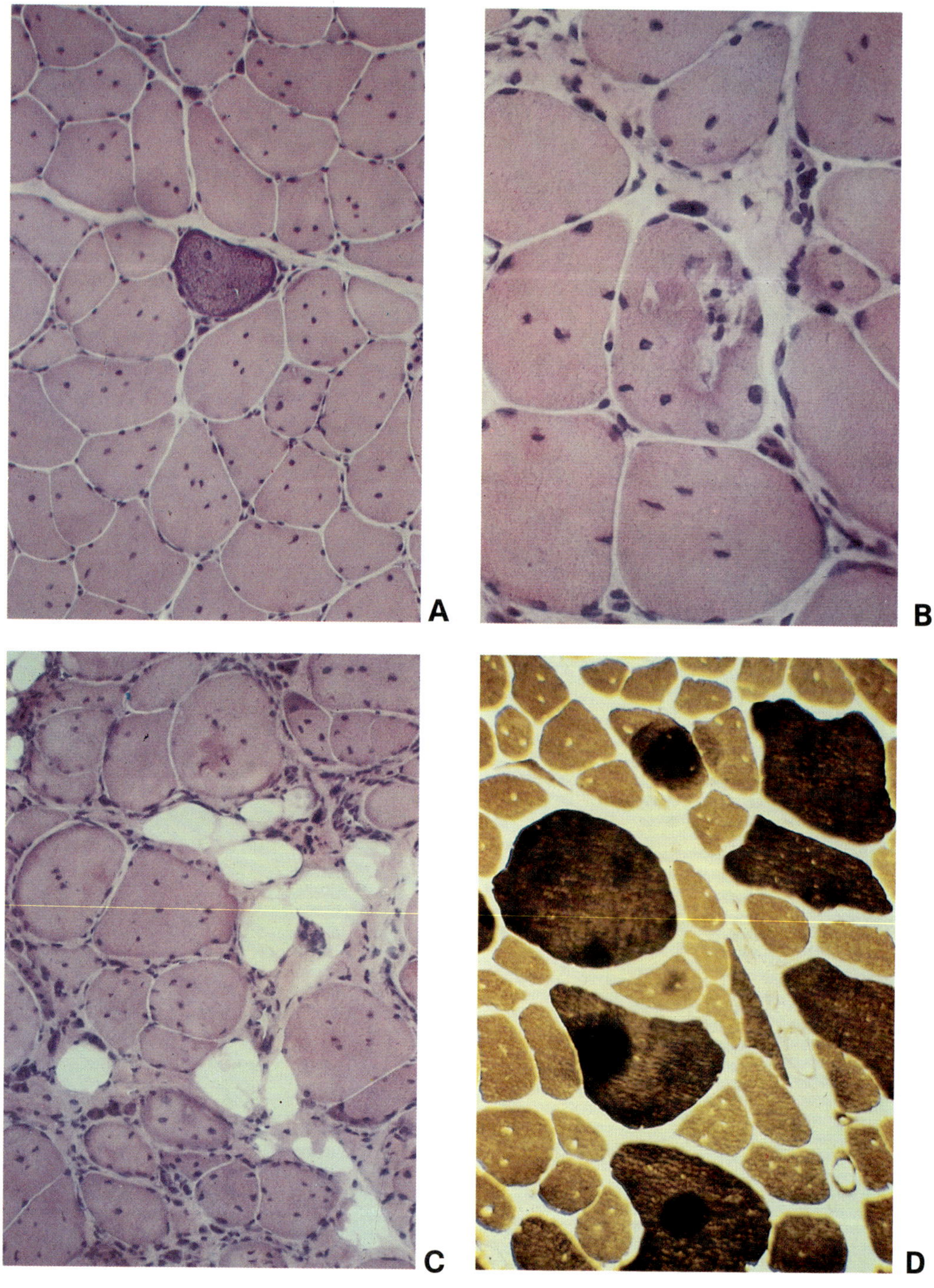

Plate 44 Muscle specimens from patients with myotonic dystrophy. **A,B,C:** H&E stain. **D:** Alkaline myofibrillar ATPase stain. Multiple internal nuclei and atrophy of ATPase type I fibers are virtually pathognomonic of myotonic dystrophy. Cytoplasmic bodies, ring fibers, basophilic, and moth-eaten fibers can also be seen commonly. Internal nuclei are represented by empty holes in **D.**

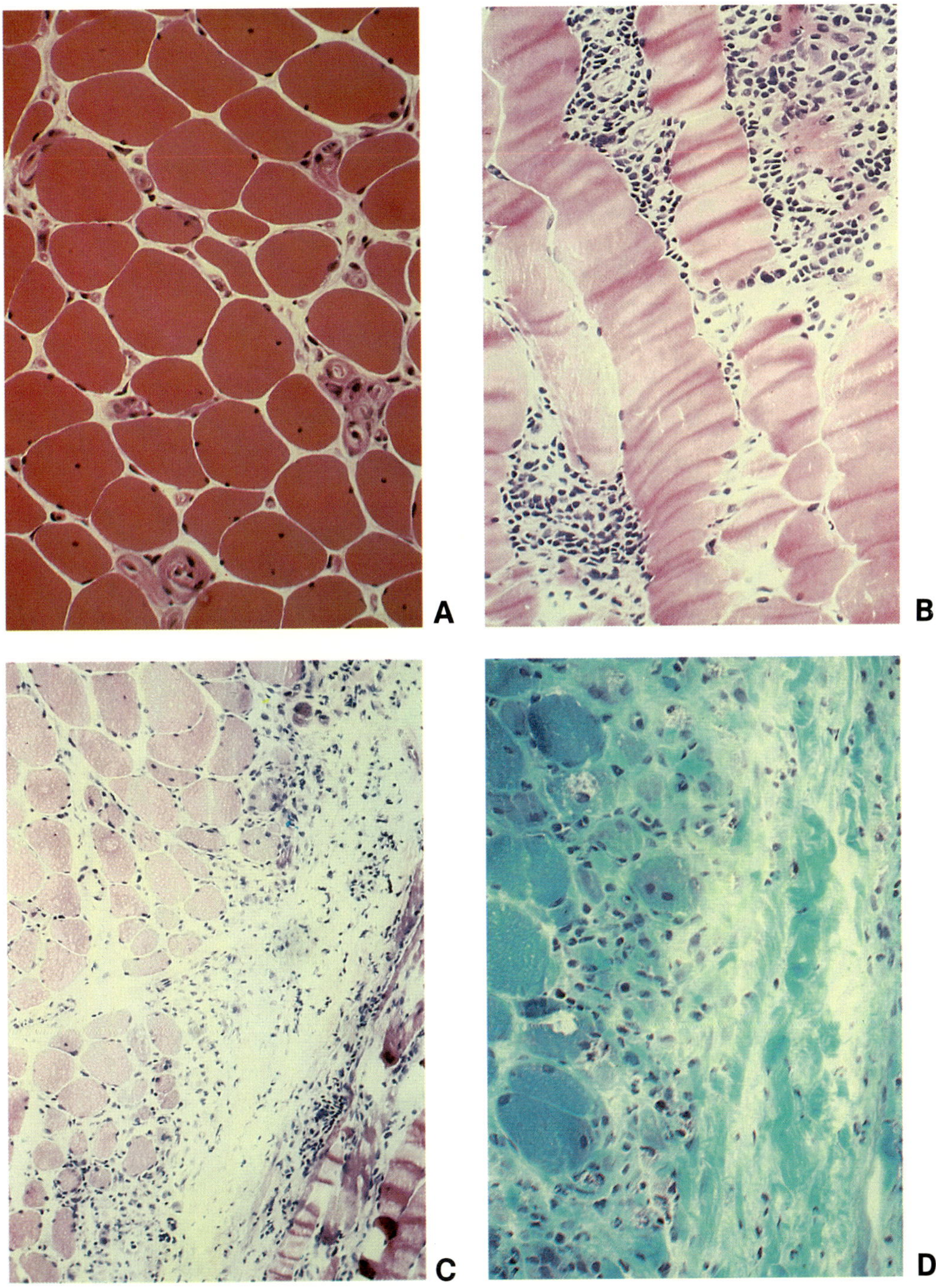

Plate 45 Cross sections of muscle specimens from patients with polymyositis. **A,B,C:** H&E stain. **D:** Modified Gomori trichrome stain. Areas of focal and widespread inflammation with inflammatory cells and muscle fiber destruction are evident. The inflammation and fiber damage is greatest at the edge of the muscle fascicle (perifascicular inflammation).

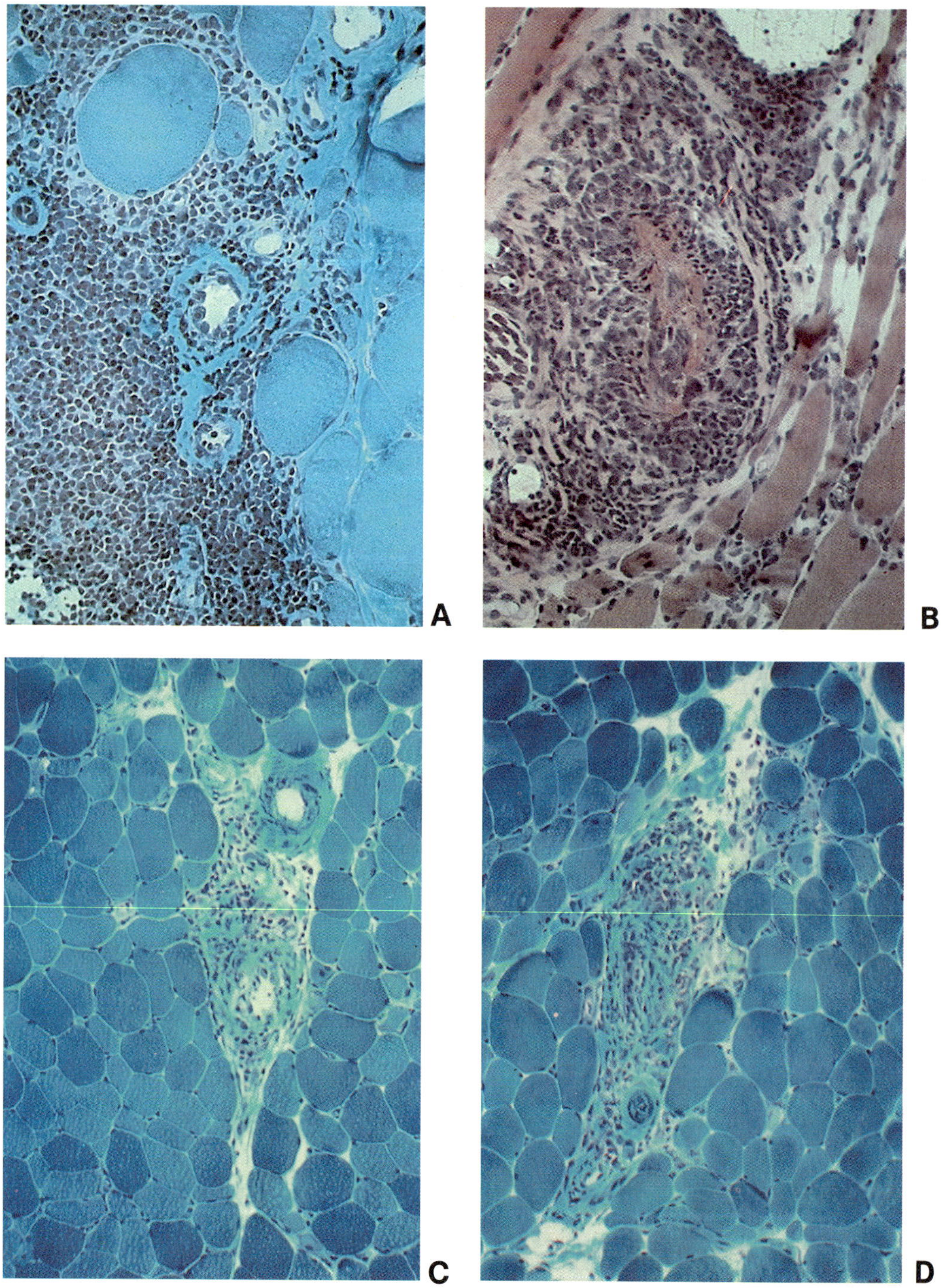

Plate 46 Cross sections of muscle specimens from patients with polymyositis. **A,C,D:** Modified Gomori trichrome stain. **B:** H&E stain. Perivascular inflammation is evident and may be quite variable in extent and severity.

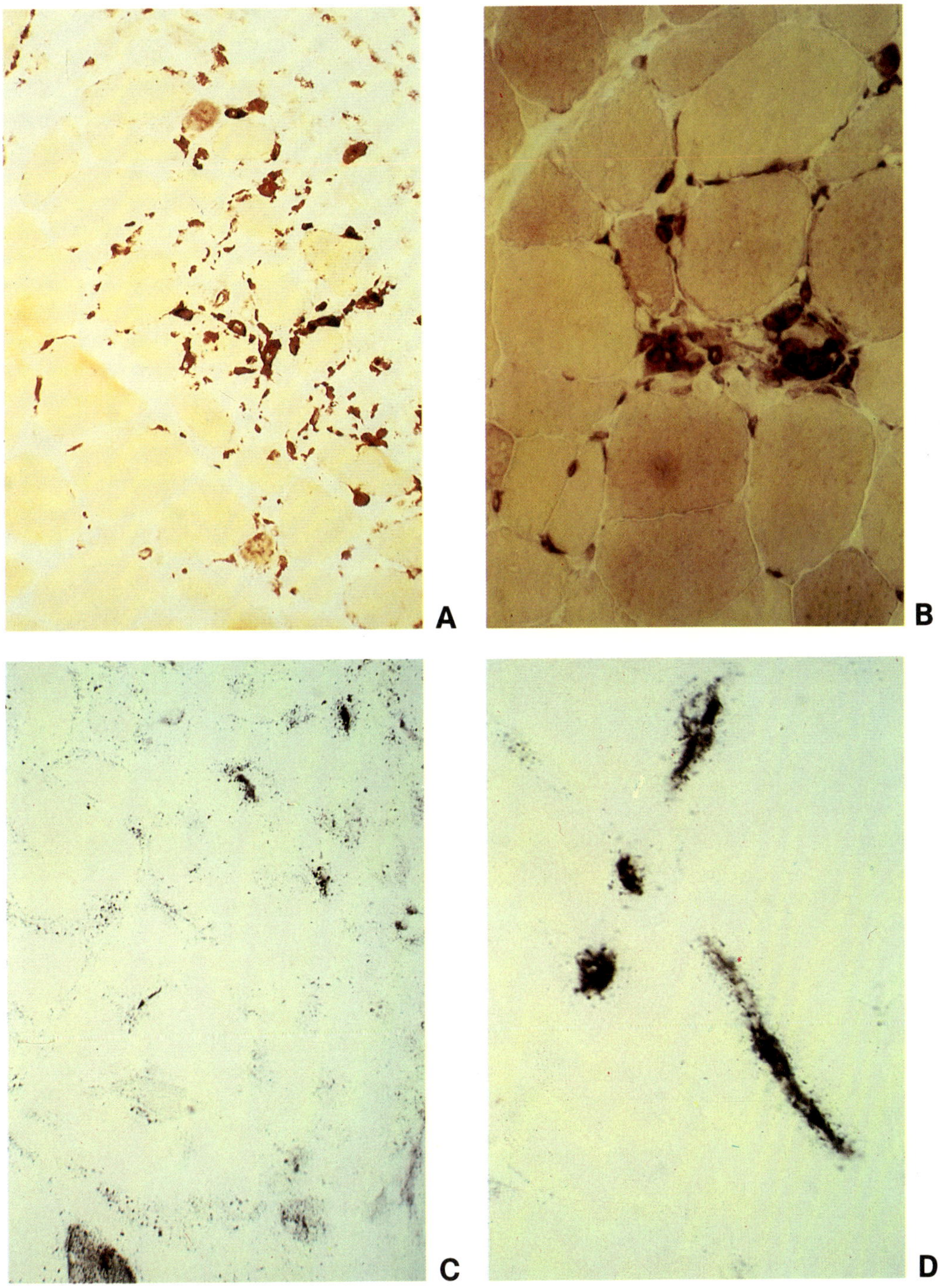

Plate 47 Cross sections of muscle specimens from patients with polymyositis. **A,B:** Nonspecific esterase stain. **C,D:** Alkaline phosphatase stain. The macrophages in the inflammatory cell infiltrates stain dark with the nonspecific esterase stain. Fibroblast proliferation in the connective tissue of the inflamed muscle stains dark with alkaline phosphatase.

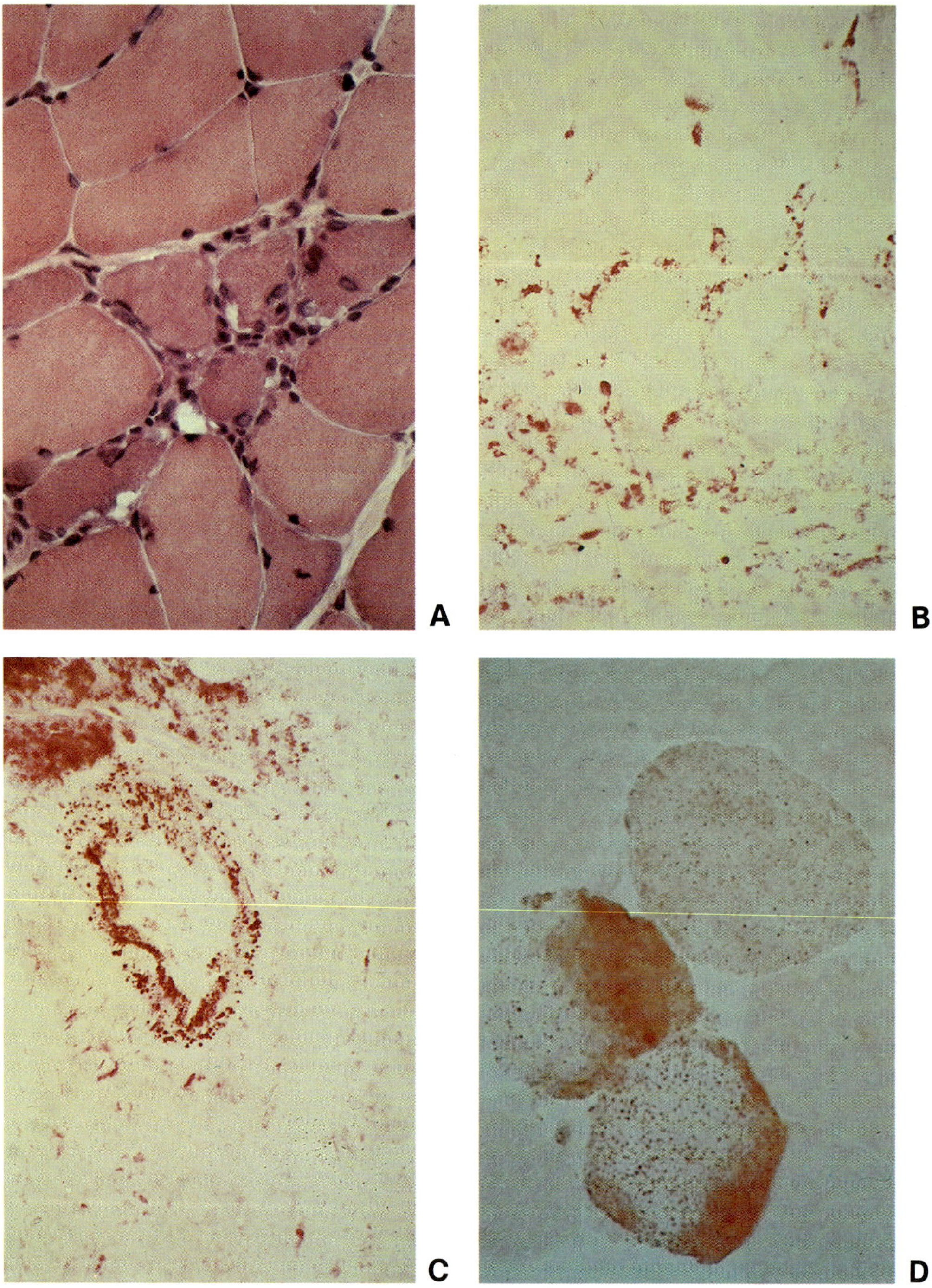

Plate 48 Cross sections of muscle specimens from patients with polymyositis. **A:** H&E stain. **B,C,D:** Alizarin red S stain. Basophilic damaged fibers are present (**A**). Calcium deposition is shown by the alizarin red S stain in damaged tissue (**B,C**); this may correlate with calcinosis evident on clinical exam or on x-ray. An individual degenerating fiber (**D**) may also show increased calcium.

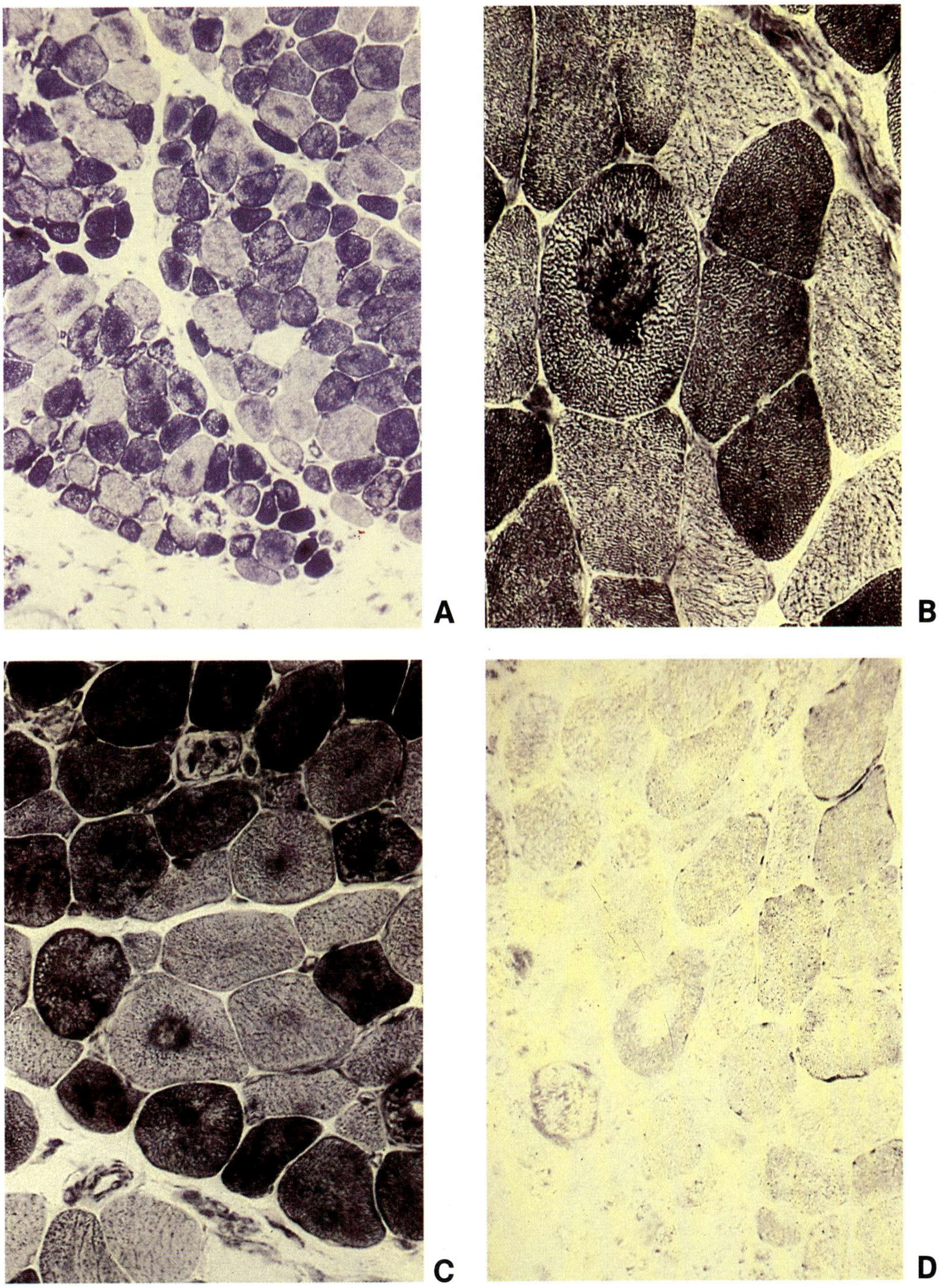

Plate 49 Cross sections of muscle specimens from patients with polymyositis. **A,B,C:** NADH-TR stain. **D:** SDH stain. Architectural changes are common in many fibers. Such alterations may be most severe at the periphery of the fascicle.

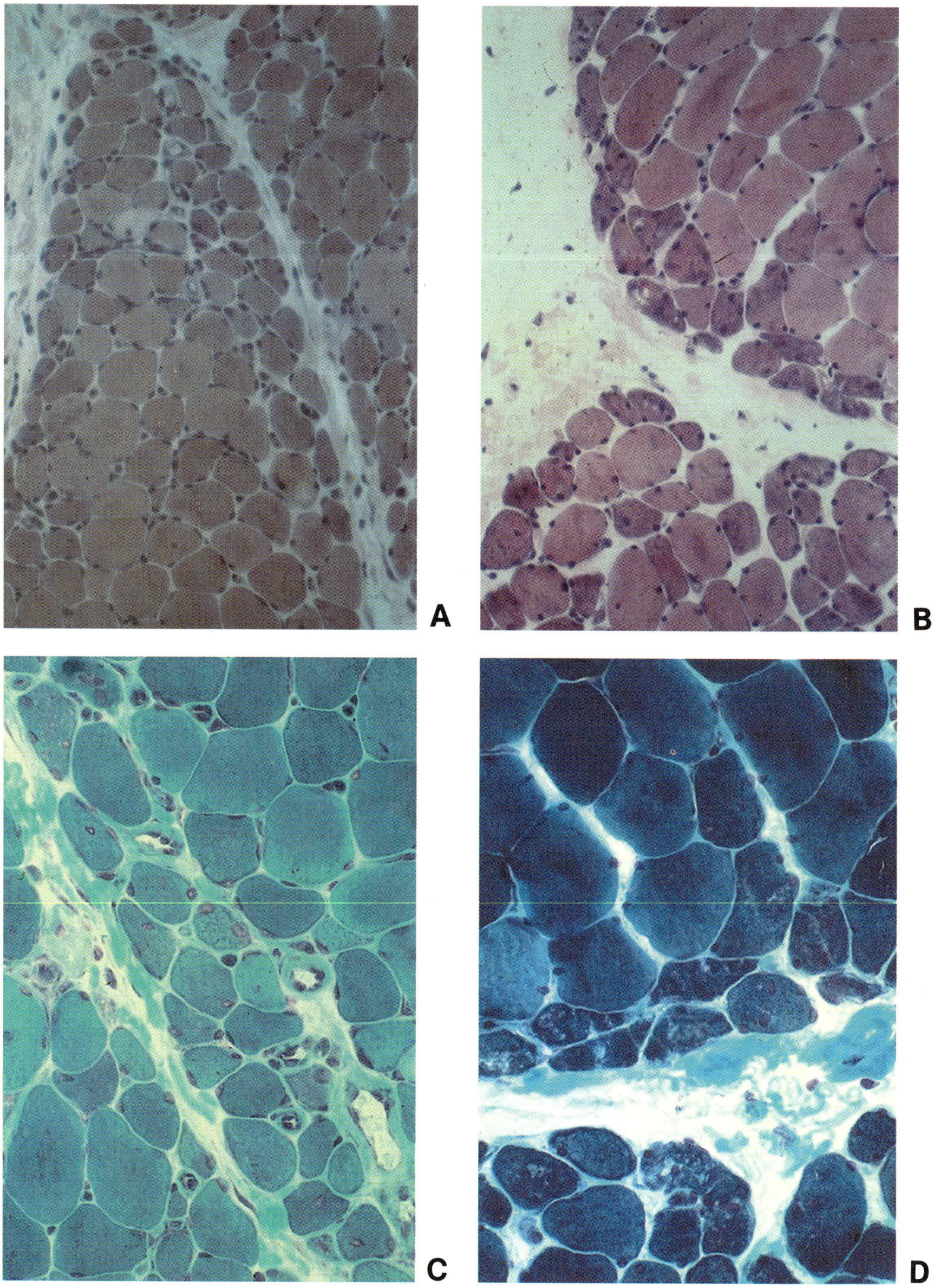

Plate 50 Cross sections of muscle specimens from patients with polymyositis. **A,B:** H&E stain. **C,D:** Modified Gomori trichrome stain. There is a perifascicular distribution of small fibers, basophilic fibers, and fibers with disturbed architecture.

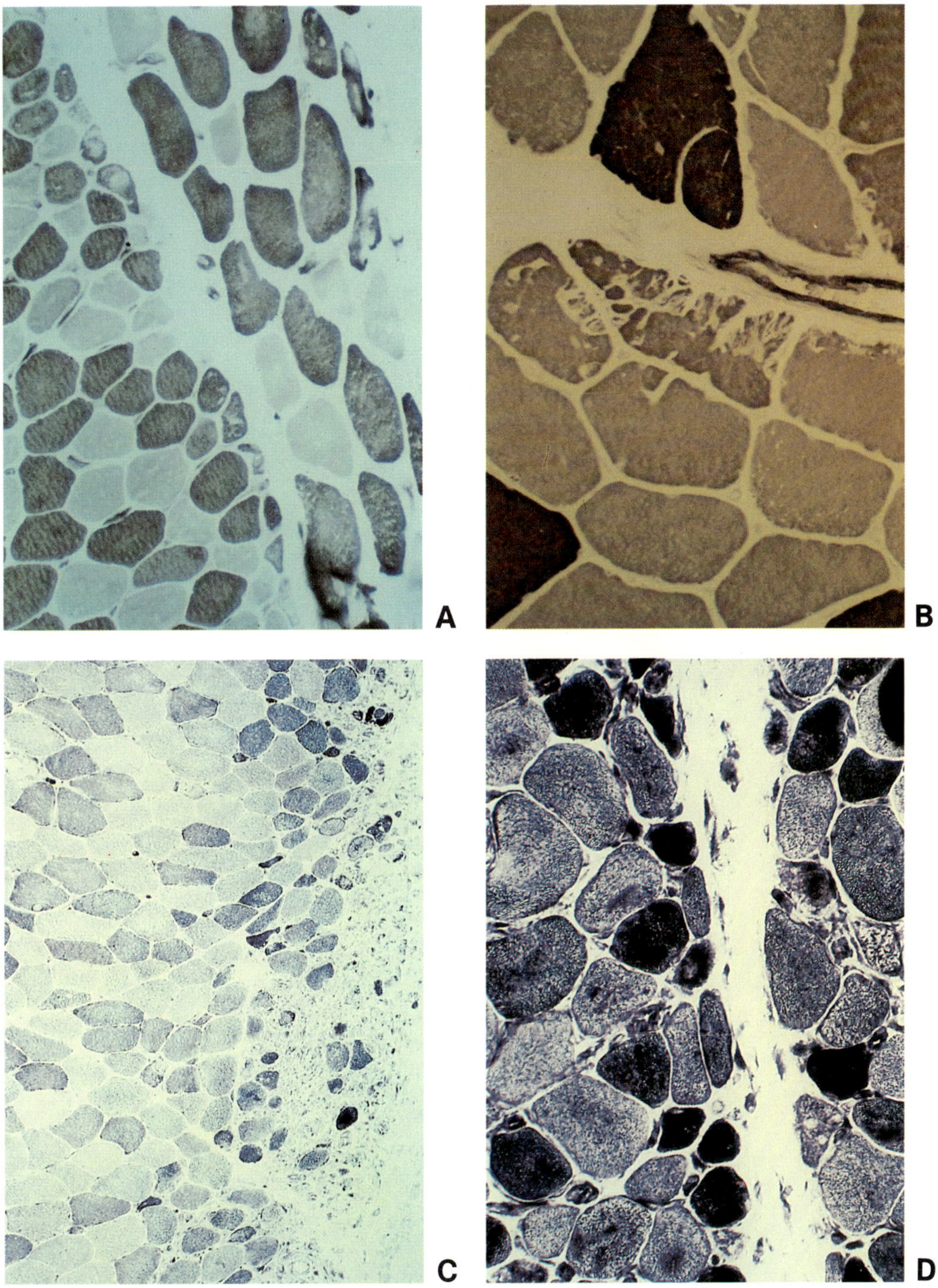

Plate 51 Cross sections of muscle specimens from patients with polymyositis. **A,B:** Alkaline myofibrillar ATPase stain. **C,D:** NADH-TR stain. Perifascicular atrophy and perifascicular architectural changes characterize polymyositis.

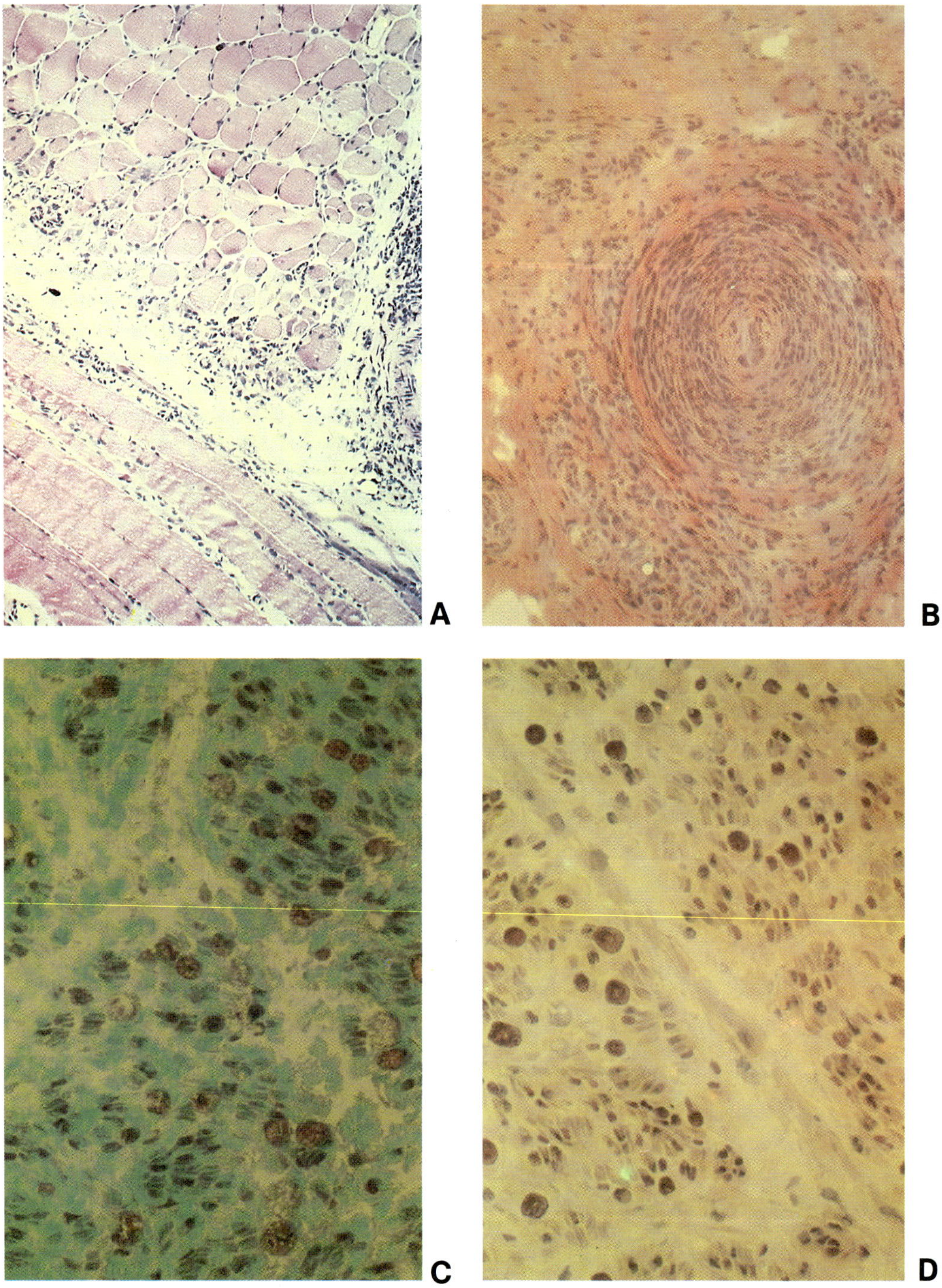

Plate 52 Cross sections of muscle (**A**) and nerve (**B,C,D**) from patients with polymyositis. **A,B,D:** H&E stain. **C:** Modified Gomori trichrome stain. Inflammation not only affects the muscle (**A**), but also is present in nerve with perivascular (vaso nervorum) inflammation (**B**) and loss of myelinated fibers (**C,D**).

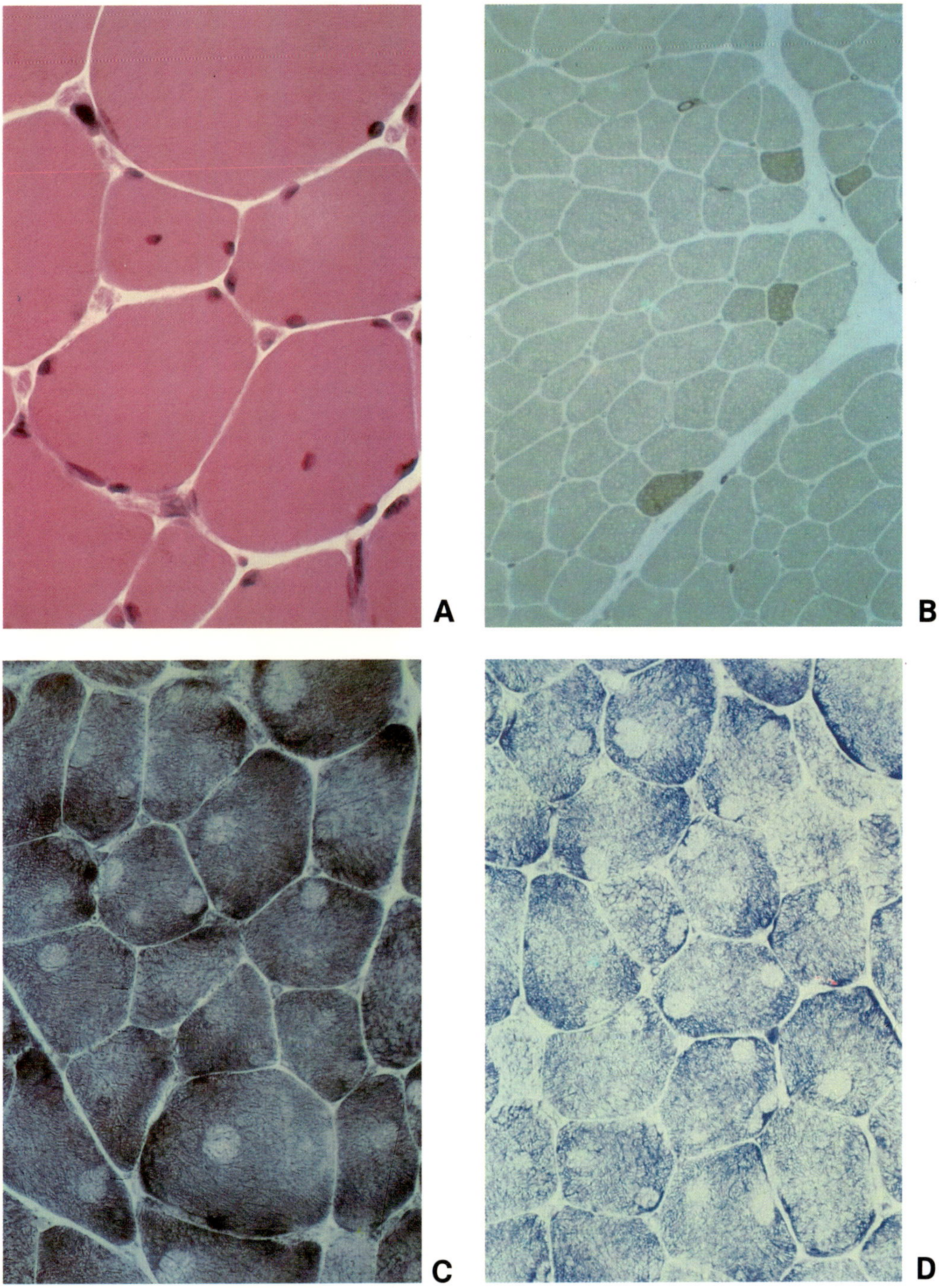

Plate 53 Cross sections of muscle specimens of central core disease. **A:** H&E stain. **B:** Alkaline myofibrillar ATPase stain. **C:** NADH-TR stain. **D:** SDH stain. In this disorder, muscle fibers are essentially unremarkable by H&E stain (**A**), but there are few or no type II fibers (**B**), and there are single target-like structures called *central cores* in all the type I fibers (**C,D**).

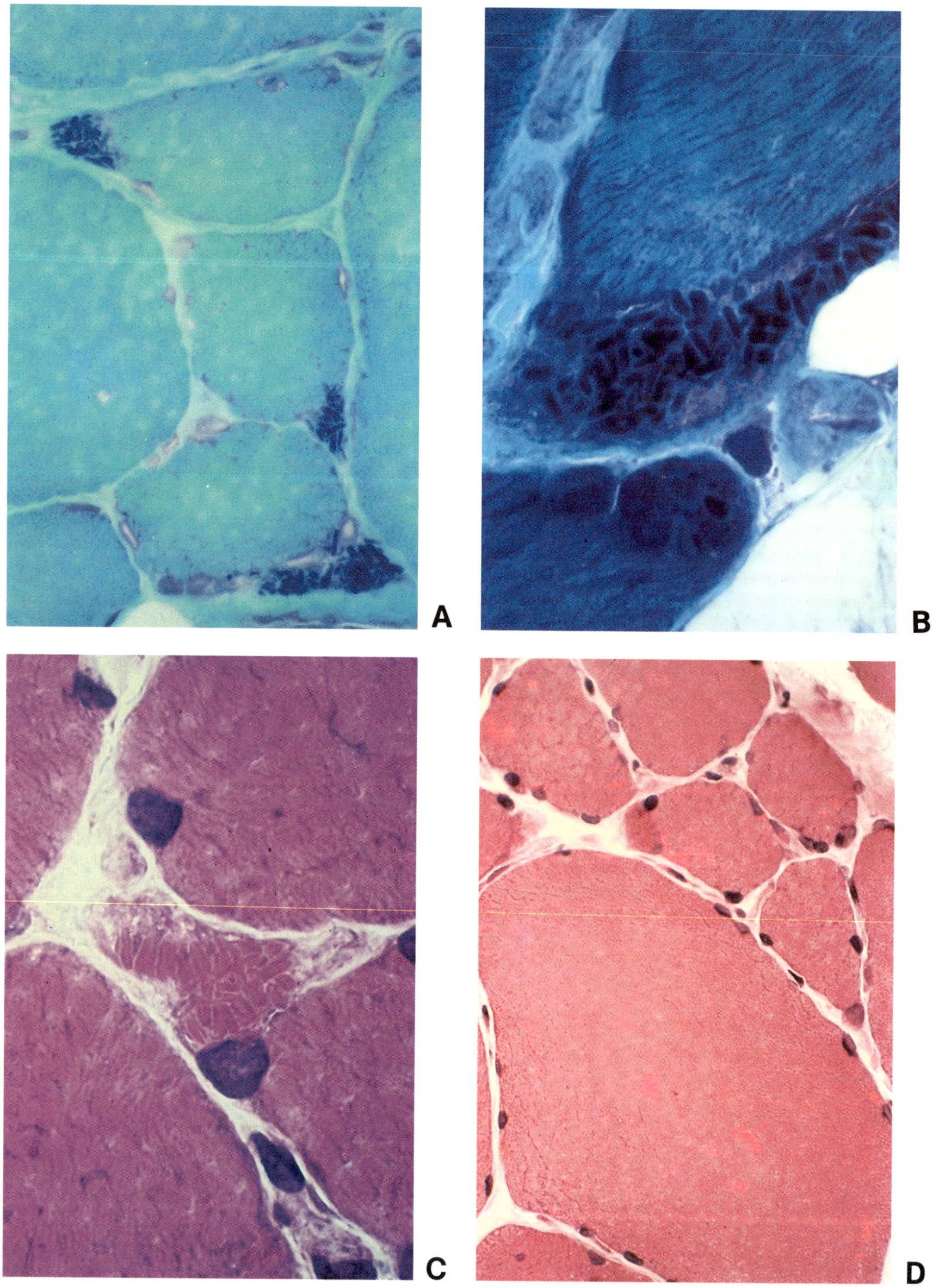

Plate 54 Cross sections of muscle specimens of rod disease (nemaline myopathy). **A,B:** Modified Gomori trichrome stain. **C,D:** H&E stain. In this disorder, type I fibers have numbers of subsarcolemmal rod-like structures seen best in the modified Gomori trichrome stain (**A,B**). The congenital form of the disorder is characterized by a virtual absence of type II muscle fibers, while the adult onset form has widespread atrophy of type II fibers and mild atrophy of type I fibers (**D**). The areas of rods are unstained by other histochemical stains.

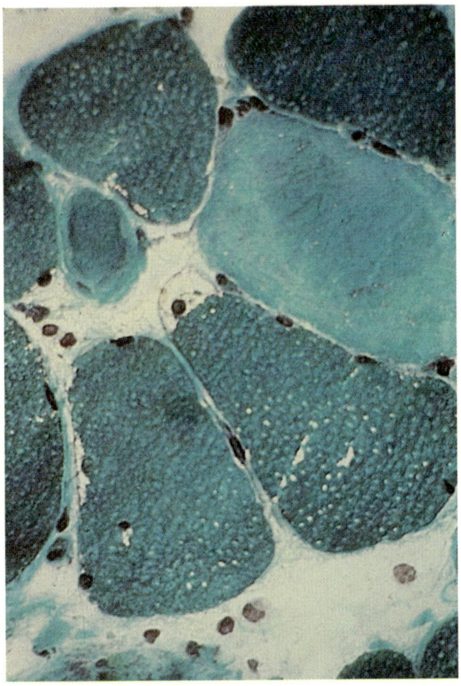
A

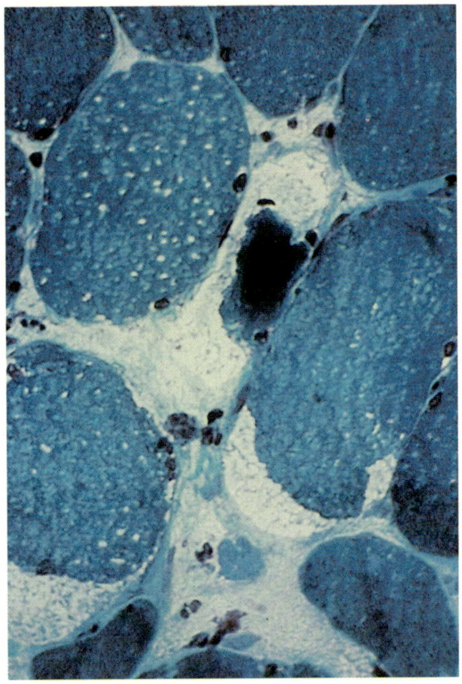
B

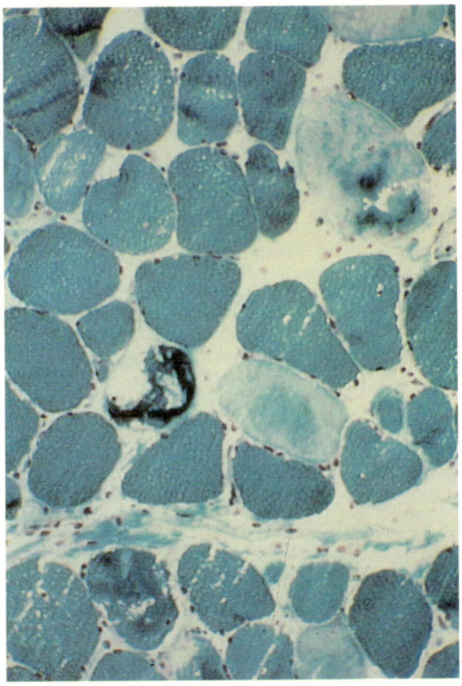
C

Plate 55 Cross sections of muscle specimens from patients with myophosphorylase deficiency (McArdle's disease). **A,B,C:** Modified Gomori trichrome stain. This disorder is a glycogen storage disease (glycogenosis type V) and unstained subsarcolemmal blebs of glycogen are visible (**A,B**). There is postexercise rhabdomyolysis (muscle cell breakdown) secondary to deficient energy production (due to defective ability to produce glucose from glycogen) (**C**).

A

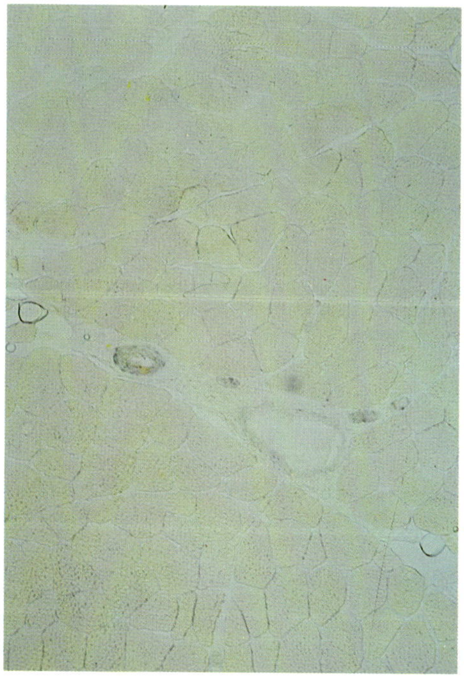

B

Plate 56 Cross sections of muscle specimens from patients with myophosphorylase deficiency (**A,B**) and normal control (**C**). **A,B,C:** Amylophosphorylase stain. The normal control specimen (**C**) shows staining of phosphorylase activity, which is dark in the type II fibers and light in the type I fibers. There is no staining of skeletal muscle fibers in patients with myophosphorylase deficiency. However, the stain is unstable and as the iodine sublimes, staining will disappear even in normal specimens. It is possible to differentiate an old stain (with iodine sublimed) from true myophosphorylase deficiency by positive amylophosphorylase staining of the smooth muscle fibers of the blood vessel walls (**B**).

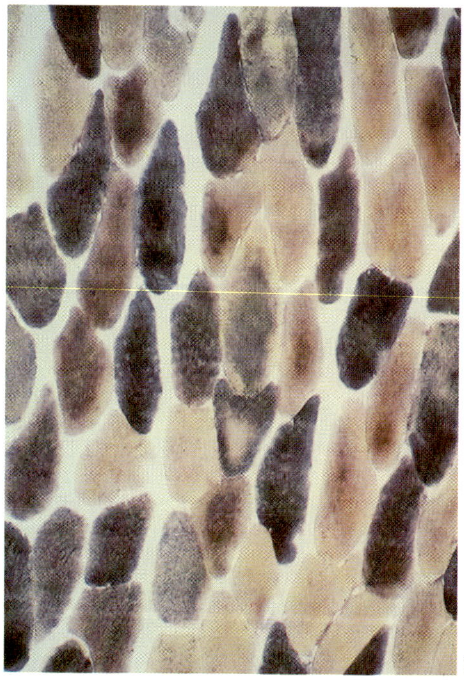

C

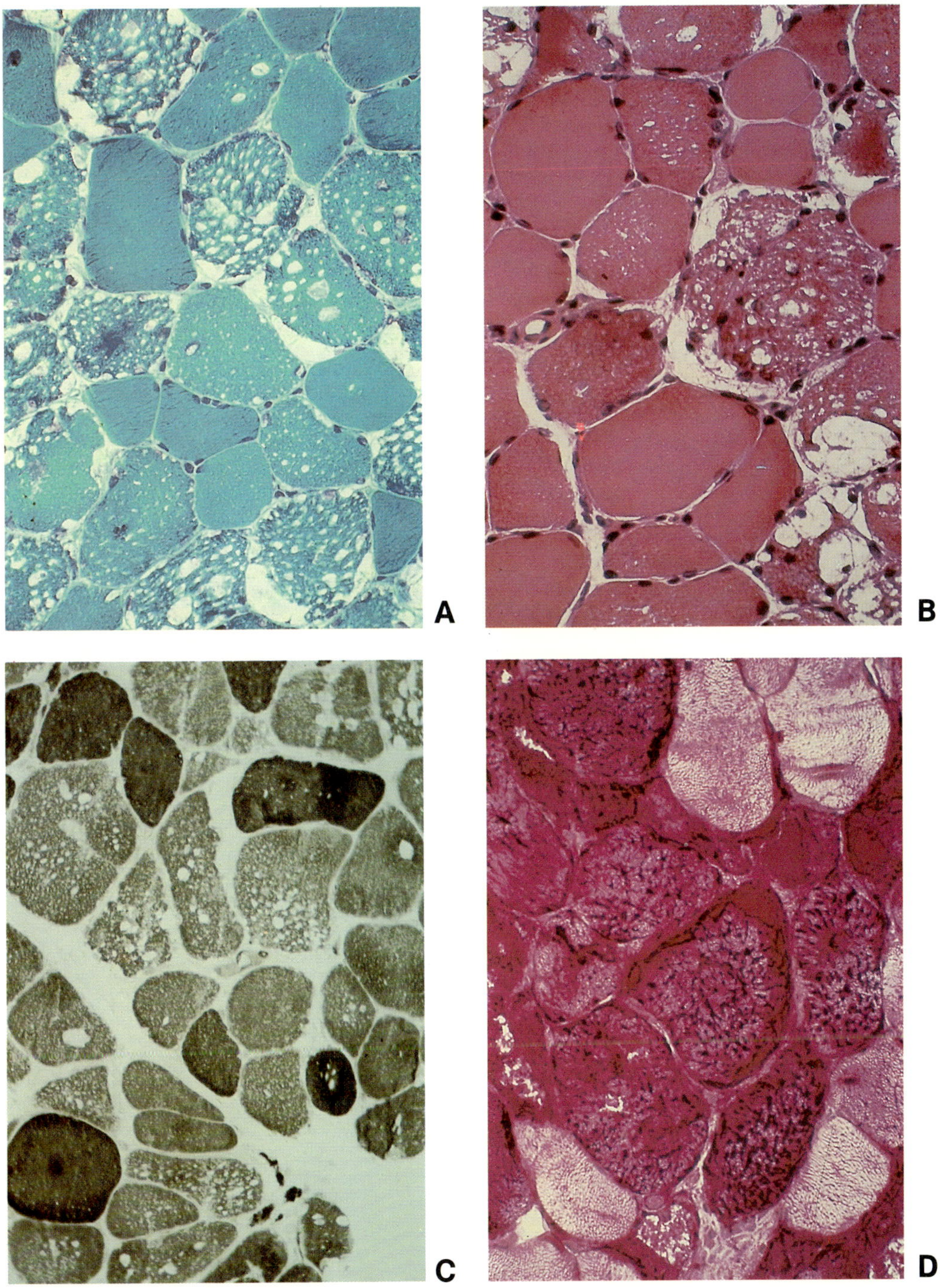

Plate 57 Cross sections of muscle specimens from acid maltase deficiency. **A:** Modified Gomori trichrome stain. **B:** H&E stain. **C:** Alkaline myofibrillar ATPase stain. **D:** Periodic acid–Schiff stain. Many of the fibers will have a "Swiss cheese" appearance with material apparent in the vacuoles (**A,B**). The vacuoles are in type I fibers only (**C**). This disorder is a glycogen storage disease (glycogenosis type II) and the vacuoles in the muscle fibers contain large quantities of glycogen (**D**).

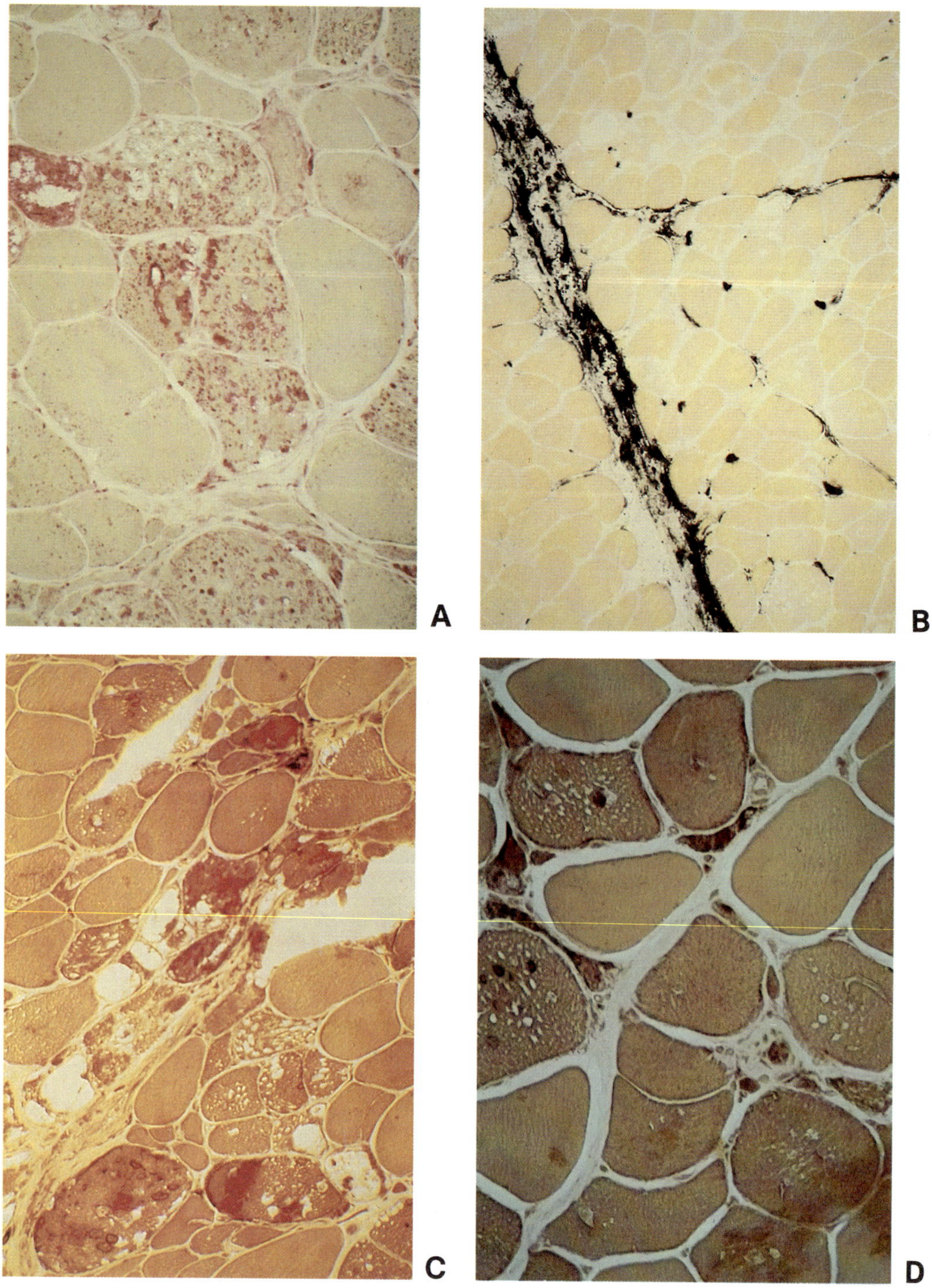

Plate 58 Cross sections of muscle specimens from acid maltase deficiency. **A:** Acid phosphatase stain. **B:** Alkaline phosphatase stain. **C,D:** Nonspecific esterase stain. This disorder is a deficiency of the lysosomal enzyme acid maltase with storage of the glycogen in enlarged lysosomes. These enlarged lysosomes also stain dark with the acid phosphatase (**A**) and nonspecific esterase stains (**C,D**). There is positive alkaline phosphatase staining of the connective tissue (**B**).

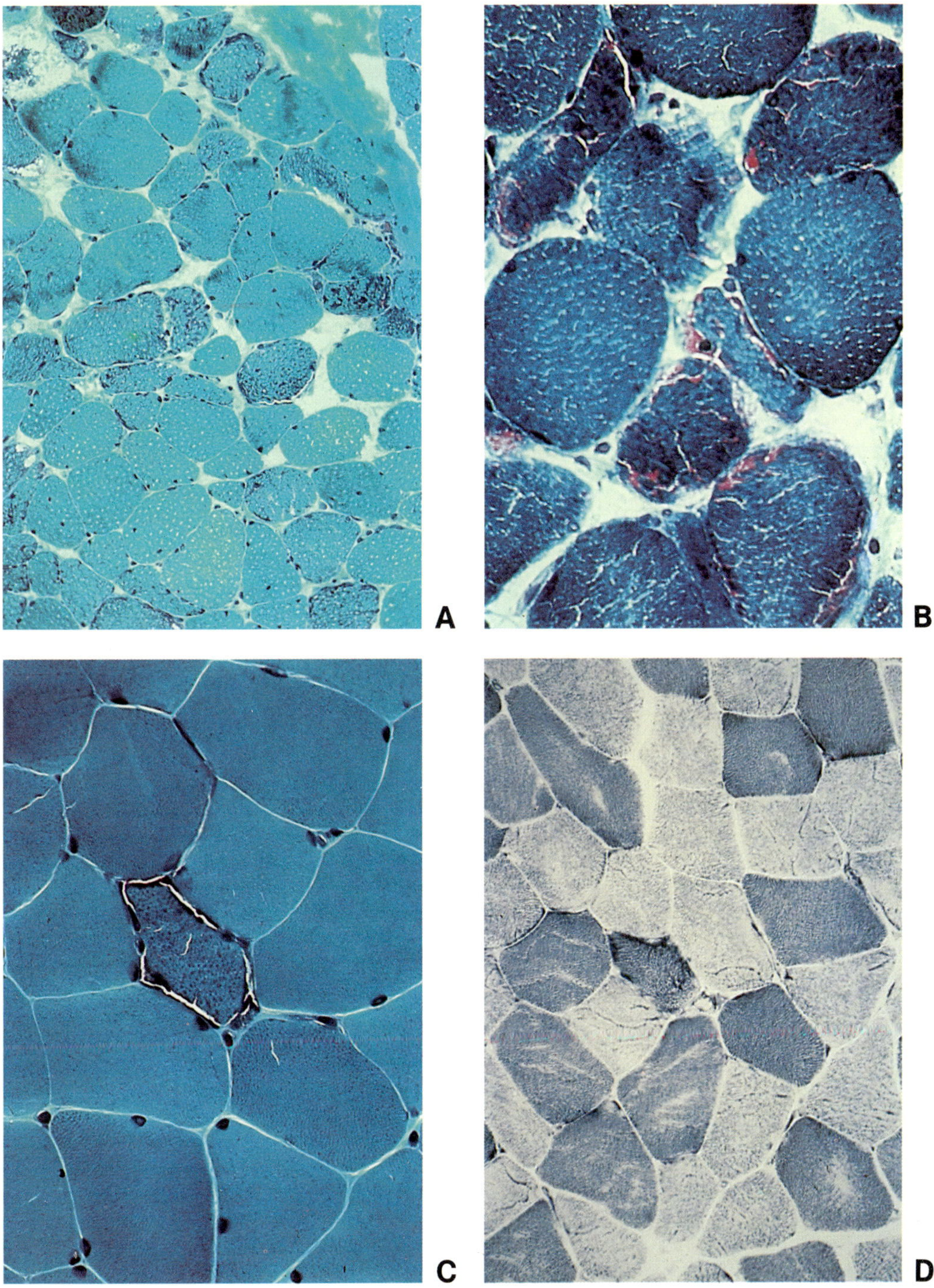

Plate 59 Cross sections of muscles showing ragged red fibers. **A,B,C:** Modified Gomori trichrome stain. **D:** NADH-TR stain. Ragged red fibers may be seen singly (**C,D**) or there may be many ragged red fibers. These fibers have a thin, reddish rim and appear cracked with the modified Gomori trichrome stain (**A,B,C**). With the NADH-TR stain, there is dark staining of the rim of the fibers and the cracks. The ragged red fibers are type I fibers.

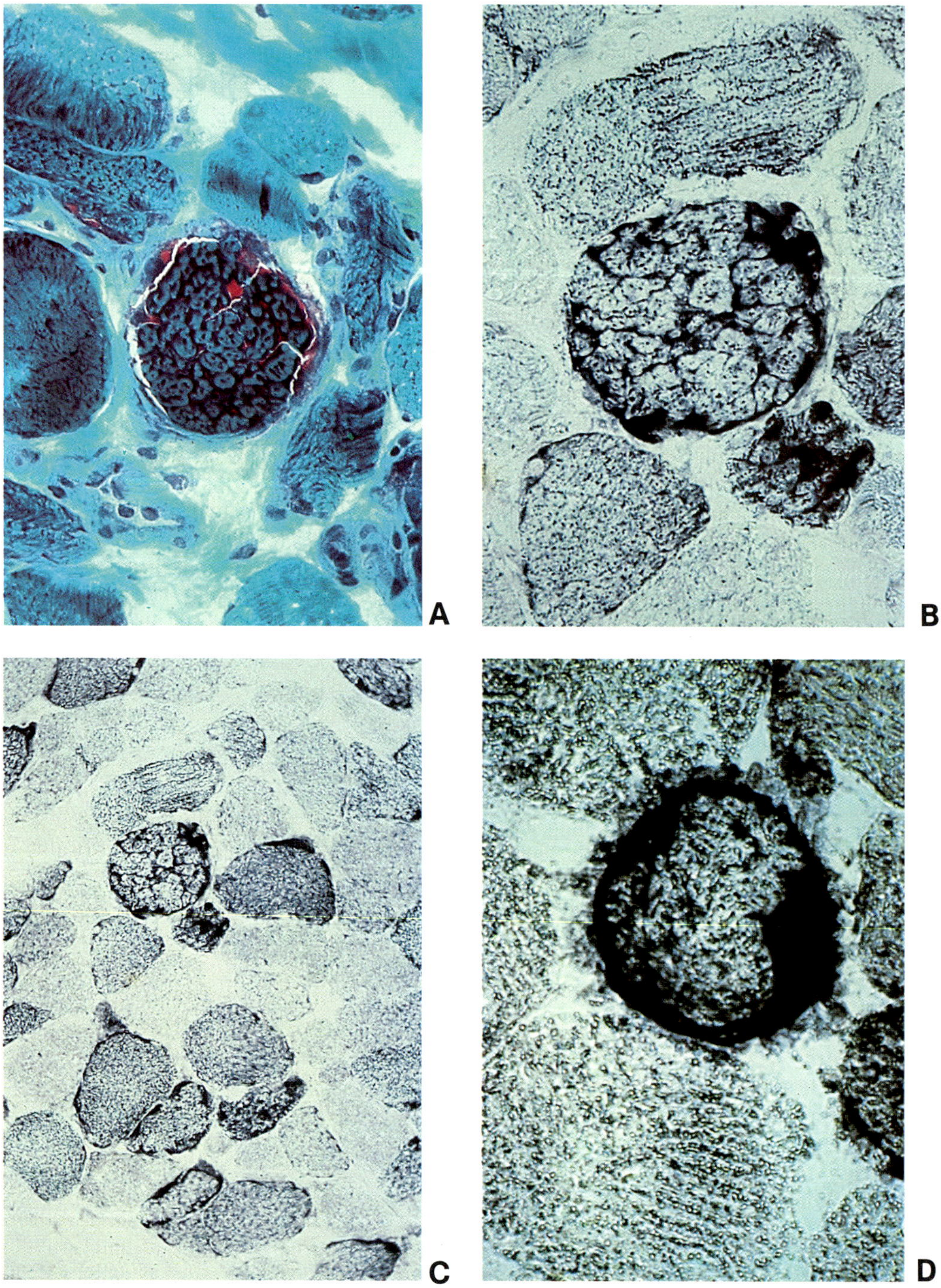

Plate 60 Cross sections of muscles showing ragged red fibers. **A:** Modified Gomori trichrome stain. **B,C,D:** SDH stain. The ragged red fibers contain abnormal collections of bizarre-shaped mitochondria (by electron microscopy). Thus, with the SDH stain for mitochondria there is ragged dark staining (**B,C,D**).

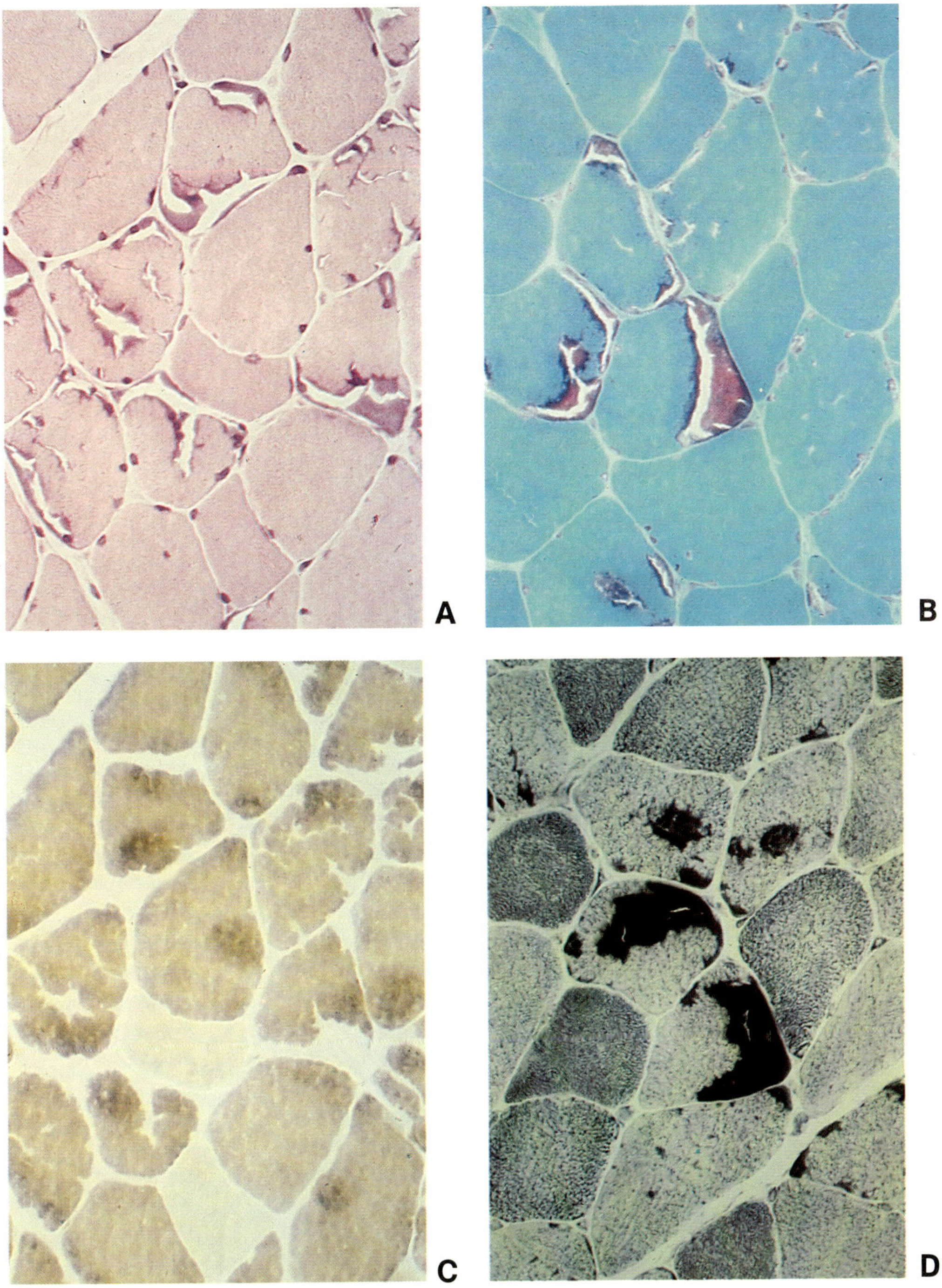

Plate 61 Cross sections of muscle specimens showing tubular aggregates (aggregates of sarcoplasmic reticulum tubules). **A:** H&E stain. **B:** Modified Gomori trichrome stain. **C:** Alkaline myofibrillar ATPase stain. **D:** NADH-TR stain. Tubular aggregates in the type II fibers (**C**) stain dark with the NADH-TR stain for sarcoplasmic reticulum (**D**), appear slightly basophilic with the H&E stain (**A**), and reddish with the modified Gomori trichrome stain (**B**).

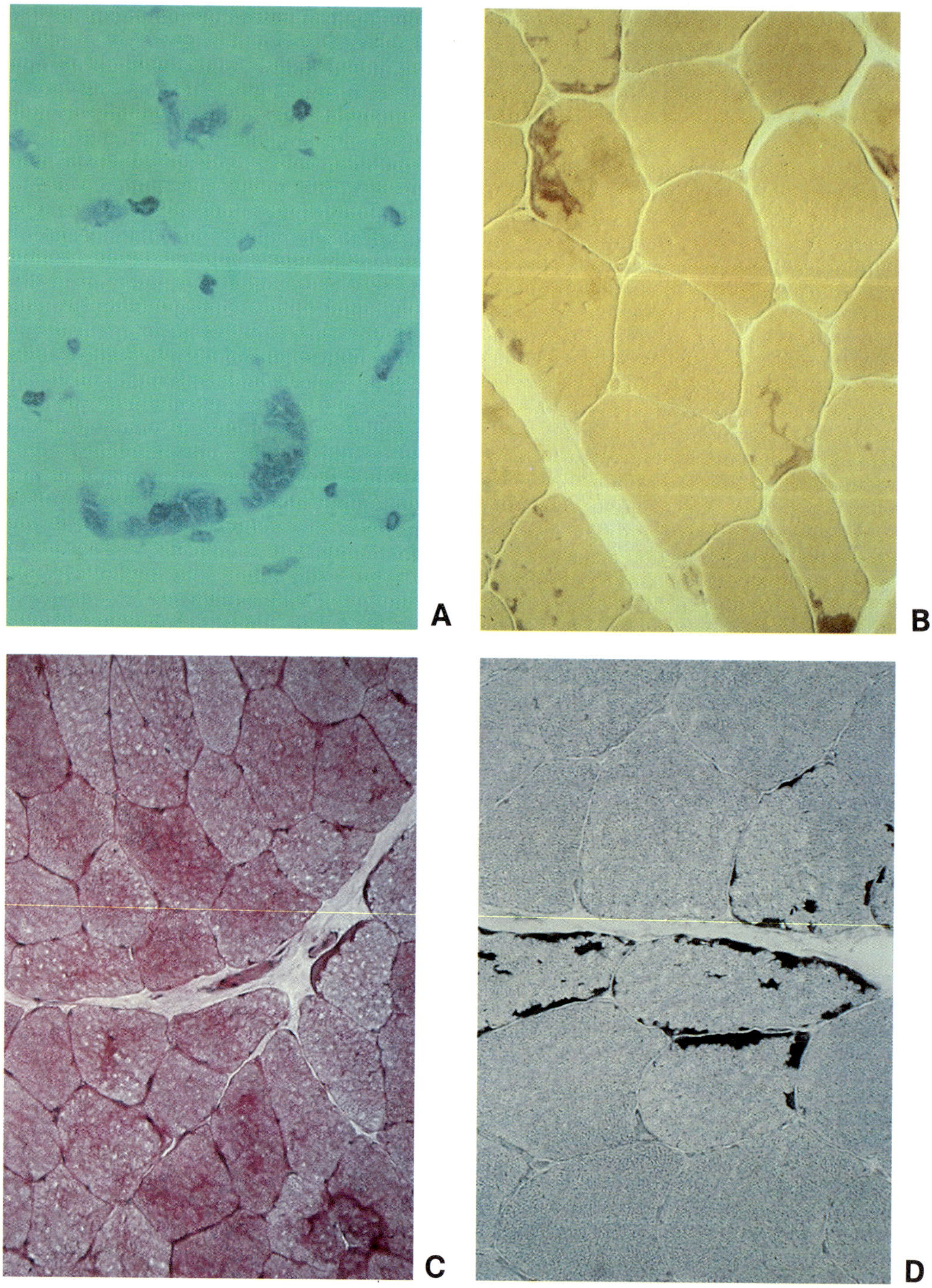

Plate 62 Cross sections of muscle specimens showing tubular aggregates. **A:** Toluidine blue stain. **B:** Nonspecific esterase stain. **C:** Periodic acid–Schiff stain. **D:** Adenylate deaminase stain. Tubular aggregates stain dark with each of these stains.

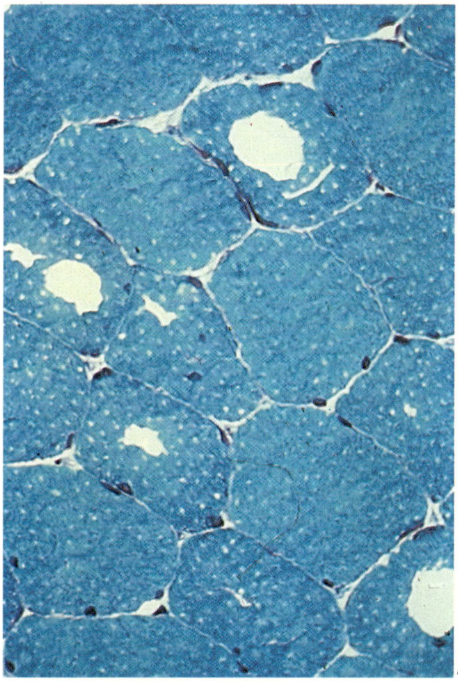

A

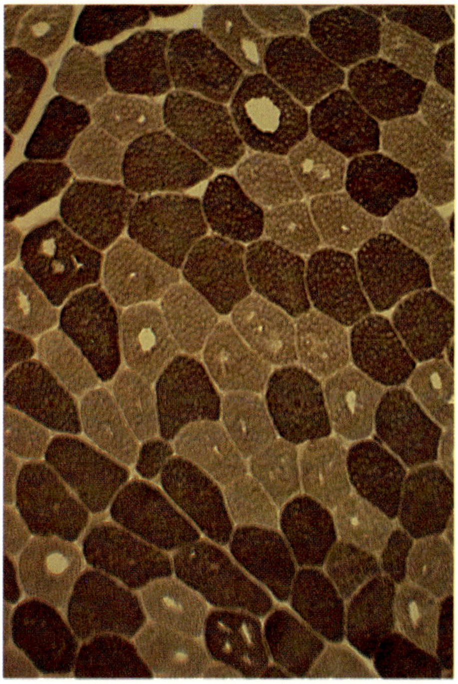

B

Plate 63 Cross sections of muscle specimens from patient with hypokalemic periodic paralysis. **A:** Modified Gomori trichrome stain. **B:** Alkaline myofibrillar ATPase stain. There are usually single clear (containing no material) vacuoles in many fibers of both fiber types. Tubular aggregates also may occasionally be seen in muscle fibers in this disorder.

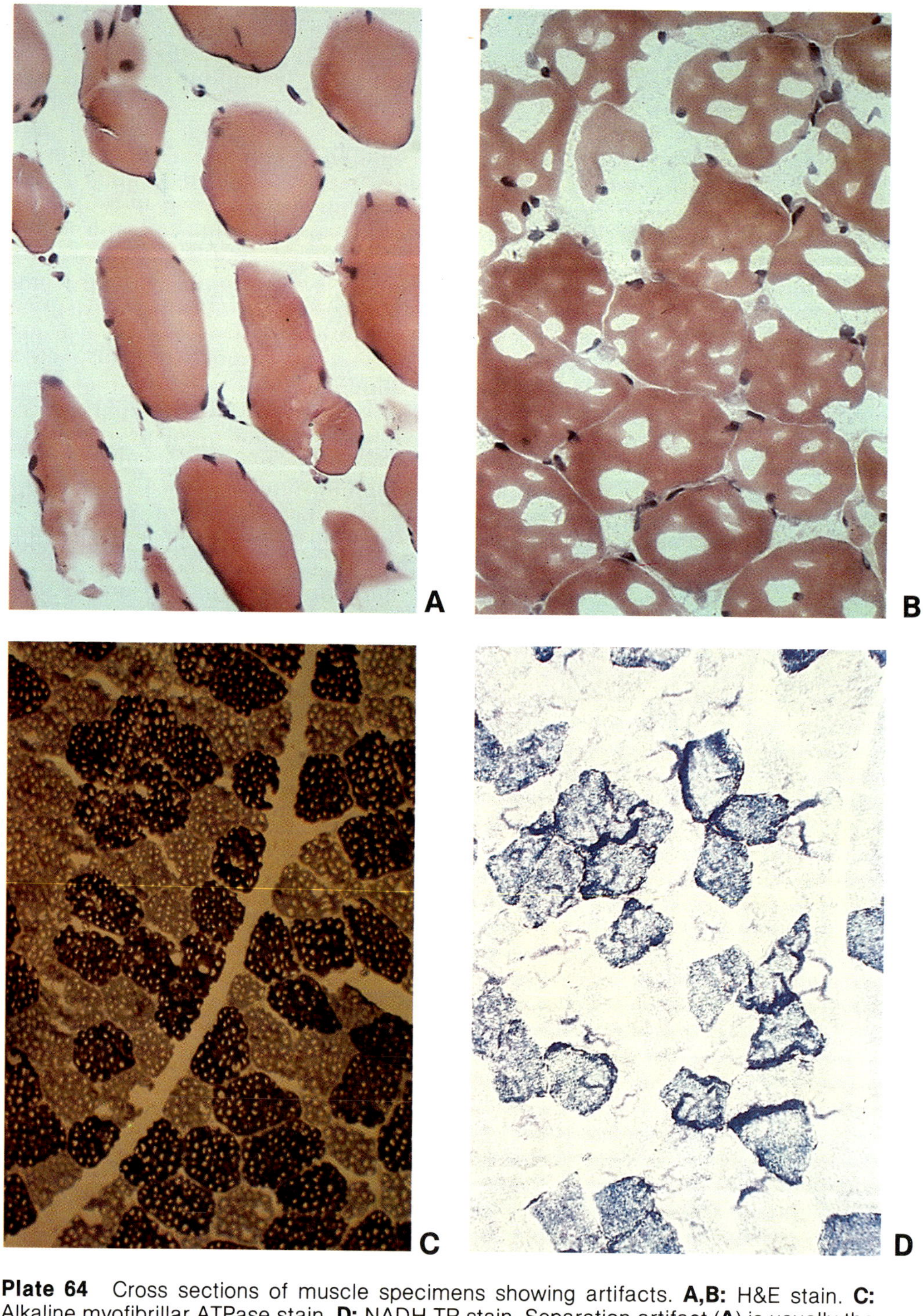

Plate 64 Cross sections of muscle specimens showing artifacts. **A,B:** H&E stain. **C:** Alkaline myofibrillar ATPase stain. **D:** NADH-TR stain. Separation artifact (**A**) is usually the result of transporting specimens on too moist a carrier. Freeze artifact appearing as mutliple holes in the specimen occurs from improper freezing techniques (**B,C**). Crinkling often occurs from cutting a specimen before it has warmed to cryostat temperature or before all the isopentane used for freezing has evaporated (**D**).

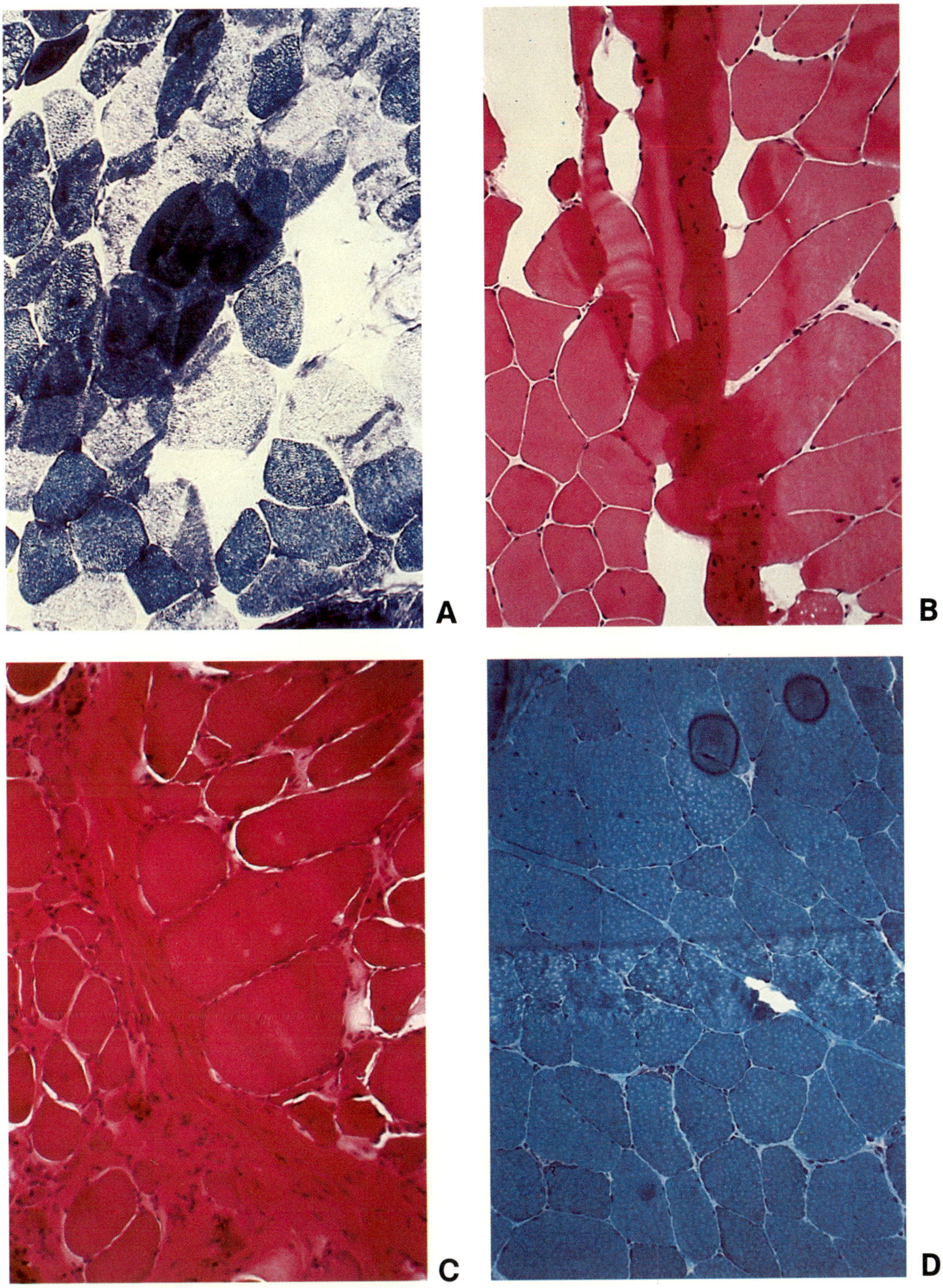

Plate 65 Cross sections of muscle specimens showing artifacts. **A:** NADH-TR stain. **B,C:** H&E stain. **D:** Modified Gomori trichrome stain. Folds may occur with sectioning (**A,B**). Alternating light and dark linear zones are the result of uneven cutting thickness of the specimen (**C**). A knife mark may appear as a dark line in the specimen (**D**).

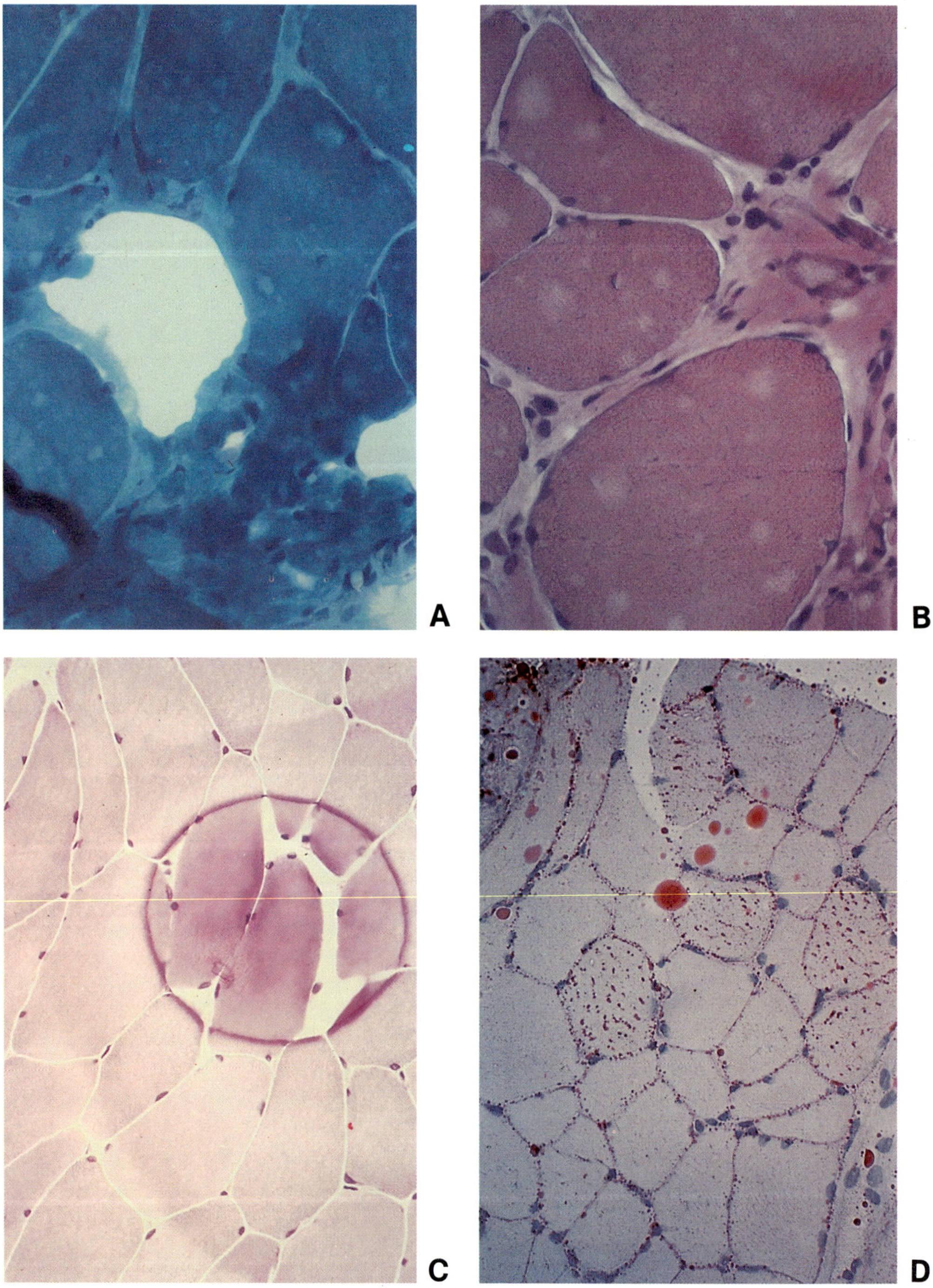

Plate 66 Cross sections of muscle specimens showing artifacts. **A:** Modified Gomori trichrome stain. **B,C:** H&E stain. **D:** Oil red O stain. Wrinkling and irregular staining may occur around areas of fat (**A**). Water droplets left during the mounting process have a fuzzy appearance and will prevent a clear focus (**B**). Air bubbles in the mounting medium have a sharp edge and can be focused through for a clear examination of the specimen (**C**). Large droplets of oil red O stain may be left on the specimen during mounting (**D**).

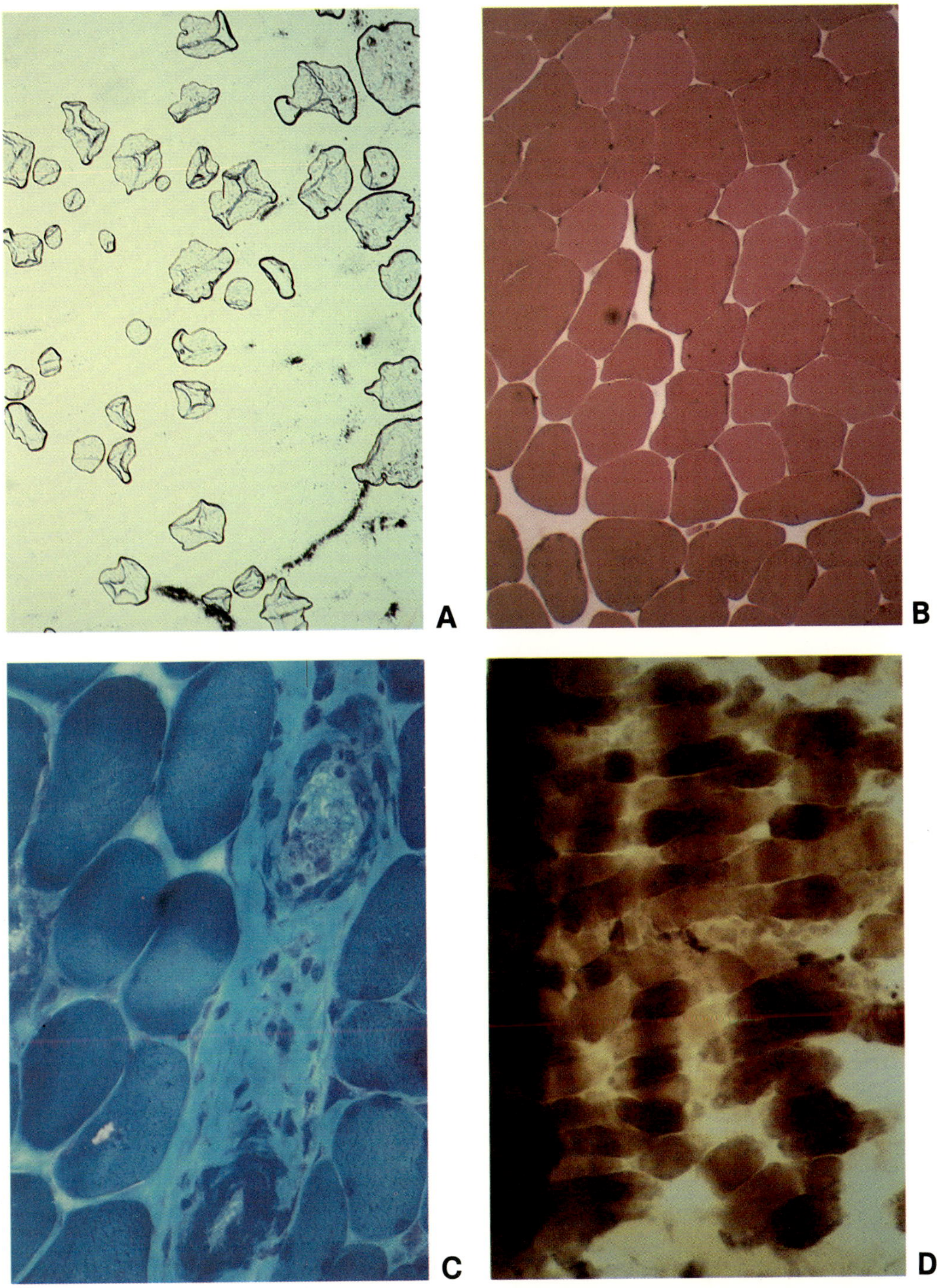

Plate 67 Cross sections of muscle specimens showing artifacts. **A:** Alkaline phosphatase stain. **B:** Alkaline myofibrillar ATPase stain. **C:** Modified Gomori trichrome stain. **D:** Non-specific esterase stain. Drying of the slide may produce crystal-like artifacts (**A**). The ATPase stain commonly fades with time, leaving no evidence of fiber differentiation. (**B**). Uneven staining is relatively common with the modified Gomori trichrome stain (**C**). Longitudinal sections will show thick and thin areas and separation due to the contraction of the fibers during the drying on the slide (after cutting) (**D**).

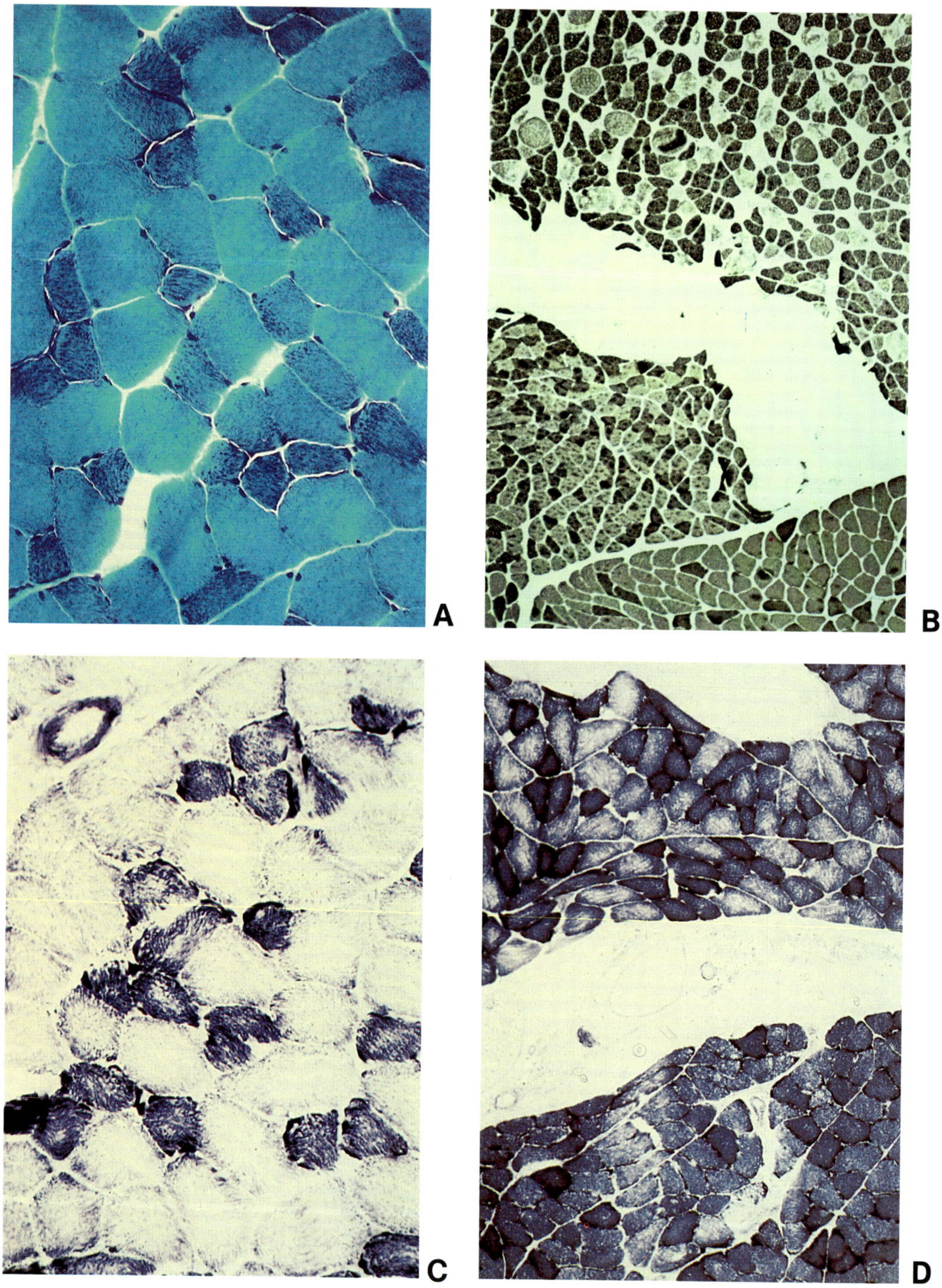

Plate 68 Cross sections of rat muscle specimens. **A:** Modified Gomori trichrome stain. **B:** Alkaline myofibrillar ATPase stain. **C,D:** NADH-TR stain. Normally, animal muscle specimens stain differently than human specimens. The muscle fibers of rats by modified Gomori trichrome stain appear to be ragged red, and by NADH-TR have dark rims and light central staining (**D**). In addition, instead of an even checkerboard distribution of ATPase fiber types, certain muscles have a predominance of one fiber type (**B,C,D**).

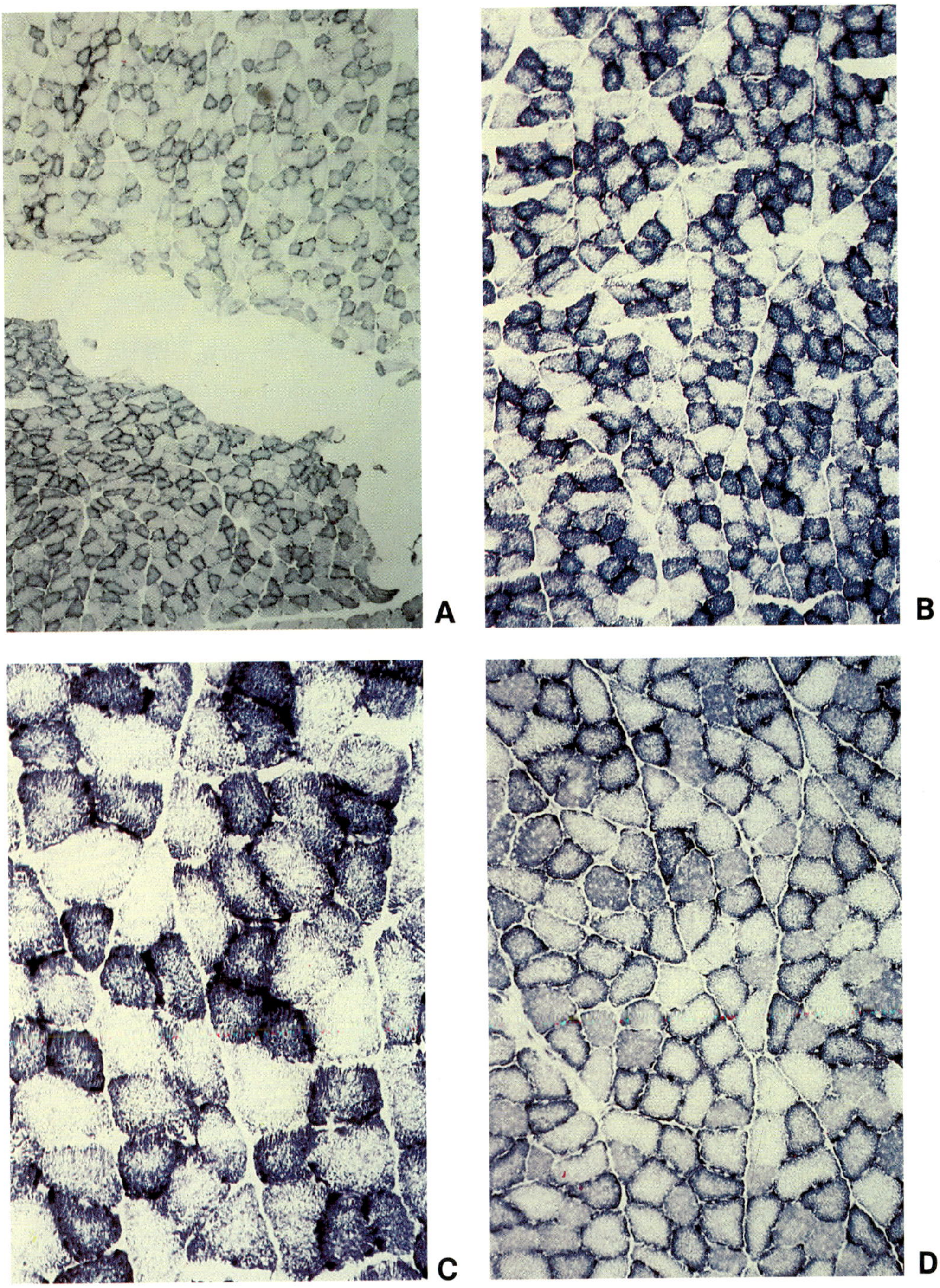

Plate 69 Cross sections of rat muscle specimens (**A,B,C**) and guinea pig muscle specimen (**D**). **A,B,C,D:** SDH stain. Note the uneven distribution of light and dark fibers (**A**), the presence of shades of intermediate staining (**B,C**), and the difference in staining between rat (**C**) and guinea pig muscle (**D**). To assess any muscle specimen from a diseased animal, the range of normal muscle staining for that particular animal and that particular muscle must first be determined.

4 SELECTED READINGS

Muscle Biopsy, Histochemical Staining, and Normal Muscle

Bancroft J: *Histochemical Techniques*. Boston, Butterworths, 1967.

Brooke MH: *A Clinician's View of Neuromuscular Diseases*. Baltimore, Williams & Wilkins, 1977.

Dubowitz V: *Muscle Disorders in Childhood*. London, WB Saunders, 1978.

Dubowitz V, Brooke MH: *Muscle Biopsy: A Modern Approach*. London, WB Saunders, 1973.

Luna LG: *Manual of Histologic Staining Methods of the Armed Forces Institute of Pathology*. New York, McGraw-Hill Inc, 1968, ed 3.

Mastaglia FL, Walton J: *Skeletal Muscle Pathology*. Edinburgh, Churchill Livingstone Inc, 1982.

Pearce AGE: *Histochemistry, Theoretical and Applied*. Edinburgh, Churchill Livingstone Inc, 1968, ed 3.

Preece A: *A Manual for Histologic Technicians*. Boston, Little, Brown & Co, 1972.

Vinken PJ, Bruyn GW: Diseases of Muscle, Part I, *Handbook of Clinical Neurology*. North Holland, Elsevier, 1979, vol 40.

Vinken PJ, Bruyn GW: Diseases of Muscle, Part II, *Handbook of Clinical Neurology*. North Holland, Elsevier, 1979, vol 41.

Walton J: *Disorders of Voluntary Muscle*. Edinburgh, Churchill Livingstone Inc, 1981, ed 4.

Denervation

Bradley WG: *Disorders of Peripheral Nerve*. Oxford, Blackwell Scientific, 1974.

Brumback RA: Failure of oral versus parenteral corticosteroids in a case of acute inflammatory polyradiculoneuropathy (Guillain-Barré syndrome). *Aust NZ J Med* 1980;10:224.

Dyck PJ, Thomas PK, Lambert EH, Bunge R: *Peripheral Neuropathy*. Philadelphia, WB Saunders, 1984, ed 2.

Fagerberg S-E: Diabetic neuropathy. A clinical and histological study on the significance of vascular affections. *Acta Med Scand* 1959;(suppl 343):5.

Haymaker W, Kernohan JW: The Landry-Guillain-Barré syndrome. *Medicine* 1949;28:59.

Lord Brain, Norris FH Jr: *The Remote Effects of Cancer on the Nervous System*. New York, Grune & Stratton, 1965.

Prineas J: Polyneuropathies of undetermined cause. *Acta Neurol Scand* 1970;(suppl 44):1.

Vinken PJ, Bruyn GW: Diseases of Nerve, Part I, *Handbook of Clinical Neurology*. North Holland, Elsevier, 1970, vol 7.
Vinken PJ, Bruyn GW: Diseases of Nerve, Part II, *Handbook of Clinical Neurology*. North Holland, Elsevier, 1970, vol 8.

Duchenne Dystrophy

Bossingham, DH, Williams E, Nichols PJR: *Severe Childhood Neuromuscular Disease: The Management of Duchenne Muscular Dystrophy and Spinal Muscular Atrophy*. London, Muscular Dystrophy Group of Great Britain, 1979.
Brooke MH: *A Clinician's View of Neuromuscular Diseases*. Baltimore, Williams & Wilkins, 1977.
Rowland LP: Pathogenesis of muscular dystrophies. *Arch Neurol* 1976;33:315.
Roy L, Gibson DA: Pseudohypertrophic muscular dystrophy and its surgical management: Review of 30 patients. *Can J Surg* 1970;13:13.
Siegel TM: *The Clinical Management of Muscle Disease*. London, Heinemann, 1977.

Fascioscapulohumeral Dystrophy

Munsat TL, Piper D: Genetically determined inflammatory myopathy with fascioscapulohumeral distribution. *Neurology* 1971;21:440.
Ricker K, Mertens HG: The differential diagnosis of the myogenic (fascio)-scapulo-peroneal syndrome. *Eur Neurol* 1968;1:275.
Ricker K, Mertens HG, Schimrigk K: The neurogenic scapuloperoneal syndrome. *Eur Neurol* 1968;1:257.

Limb-Girdle Dystrophy

Bacon PA, Smith B: Familial muscular dystrophy of late onset. *J Neurol Neurosurg Psychiatry* 1971;34:93.
Corsi A, Gentili C, Todesco CV: The relationship of neuropausal muscular dystrophy to other diseases of muscle. A study of 17 cases. *J Neurol Sci* 1965;2:397.

Myotonic Dystrophy

Brumback RA, Carlson KM, Wilson H, et al: Myotonic dystrophy as a disease of abnormal membrane receptors: An hypothesis of pathophysiology and a new approach to treatment. *Med Hypotheses* 1981;7:1059.
Caughey JE, Myrianthopouls NC: *Dystrophia Myotonica and Related Disorders*. Springfield, Thomas, 1963.
Harper PS: *Myotonic Dystrophy*. London, WB Saunders, 1979.

Polymyositis

Banker BQ: Dermatomyositis of childhood. Ultrastructural alterations of muscle and intramuscular blood vessels. *J Neuropathol Exp Neurol* 1975;34:46.
Barnes BE: Dermatomyositis and malignancy. A review of the literature. *Ann Intern Med* 1976;84:68.

DeVere R, Bradley WG: Polymyositis: Its presentation, mobidity, and mortality. *Brain* 1975;98:637.
Pearson CM, Bohan A: The spectrum of polymyositis and dermatomyositis. *Med Clin North Am* 1977;61:439.
Walton JN, Adams RD: *Polymyositis*. Edinburgh, Churchill Livingstone, 1958.

Central Core Disease

Armstrong RM, Koenigsberger R, Mellinger J, et al: Central core disease with congenital hip dislocation: Study of two families. *Neurology* 1971;21:369.
Denborough MA, Dennett X, Anderson RM: Central-core disease and malignant hyperpyrexia. *Br Med J* 1973;1:272.
Shy GM, Magee KR: A new congenital non-progressive myopathy. *Brain* 1956;79:610.

Rod Disease

Brooke MH, Carroll JE, Ringel SP: Congenital hypotonia revisited. *Muscle Nerve* 1979;2:84.
Brumback RA, Borge AF, Leech RW: Postpoliomyelitis amyotrophy with rod (nemaline) bodies in skeletal muscle. *Trans Am Neurol Assoc* 1980;105:430.
Shy GM, Engel WK, Somers JE, et al: Nemaline myopathy: A new congenital myopathy. *Brain* 1963;86:793.

Myophosphorylase Deficiency

Brumback RA: Iodoacetate inhibition of glyceraldehyde-3-phosphate dehydrogenase as a model of human myophosphorylase deficiency (McArdle's disease) and phosphofructokinase deficiency (Tarui's disease). *J Neurol Sci* 1980;48:383.
Huijng F: Glycogen metabolism and glycogen storage diseases. *Physiol Rev* 1975;55:609.
Pernow BB, Havel RJ, Jennings DB: The second-wind phenomenon in McArdle's syndrome. *Acta Med Scand [Suppl]* 1967;472:294.
Rowland LP: Myoglobinuria, 1984. *Can J Neurol Sci* 1984;11:1.
Rowland LP, Lovelace RE, Schotland DL, et al: The clinical diagnosis of McArdle's disease. Identification of another family with deficiency of muscle phosphorylase. *Neurology* 1966;16:93.

Acid Maltase Deficiency

DiMauro S, Stern LZ, Mehler M, et al: Adult-onset acid maltase deficiency: A postmortem study. *Muscle Nerve* 1978;1:27.
Engel AG: Acid maltase deficiency in adults: Studies in four cases of a syndrome which may mimic muscular dystrophy or other myopathies. *Brain* 1970;93:599.
Swaiman KF, Kennedy WR, Sauls HS: Late infantile acid maltase deficiency. *Arch Neurol* 1968;18:642.

Ragged Red Fiber Disease

Berenberg RA, Pellock JM, DiMauro S, et al: Lumping or splitting? "Ophthalmoplegia plus" or Kearns-Sayre syndrome? *Ann Neurol* 1977;1:37.

Olson W, Engel WK, Walsh GO, et al: Oculocraniosomatic neuromuscular disease with 'ragged-red' fibers. *Arch Neurol* 1972;26:193.

Tubular Aggregate Myopathy

Brumback RA, Staton RD, Susag ME: Exercise-induced pain, stiffness, and tubular aggregation in skeletal muscle. *J Neurol Neurosurg Psychiatry* 1981; 44:250.

Lewis PD, Pallis C, Pearse AGE: Myopathy with tubular aggregates. *J Neurol Sci* 1971;13:381.

Periodic Paralysis

Ionasescu V, Schochet SS, Powers JM, et al: Hypokalemic periodic paralysis. *J Neurol Sci* 1974;21:419.

Shy GM, Wanko T, Rowley PT, et al: Studies in familial periodic paralysis. *Exp Neurol* 1961;3:53.

Walter GF, Auböck L: Ultrastructural changes of muscle during attacks of adynamia episodica hereditaria. *Neuropathol Appl Neurobiol* 1980;6:313.

APPENDIX

CHEMICALS USED IN MUSCLE HISTOCHEMISTRY

Absolute alcohol (ethanol)
Acetone
Acid fuchsin
Adenosine monophosphate (AMP)
Adenosine-5-triphosphate (ATP)
5′-Adenylylimidodiphosphate (AMP-PNP)
Alizarin red S
Ammonium sulfide

Basic fuchsin

Calcium chloride ($CaCl_2$)
Chromotrope 2R
Cobaltous chloride ($CoCl_2$)

Diastase, malt
N,N-Dimethylformamide
Distilled water
Dithiothreitol
Dry ice

EDTA (disodium salt)
Eosin Y
Ethanol, 95%

Fast blue RR salt
Fast green FCF

Gelatin
Giemsa's stain
Glacial acetic acid
Glucose-1-phosphate, potassium salt
Glutaraldehyde
Glyceraldehyde-3-phosphate
Glycerine
α-Glycerophosphate
Glycine
Glycogen, shellfish
5′-Guanylylimidodiphosphate (GMP-PNP)
Gum tragacanth

Hematoxylin crystals
Hydrochloric acid (HCl)

Iodine
Isopentane (2-methylbutane)
Isopropanol (isopropyl alcohol)

Lactic acid
Liquid nitrogen
Lithium carbonate

Magnesium sulfate ($MgSO_4$)
Menadione
Mercuric oxide, red
Methyl alcohol (methanol)
Methyl green

Naphthol AS-BI phosphate, sodium salt
α-Naphthyl acetate
β-Nicotinamide adenine dinucleotide, oxidized form (NAD)
β-Nicotinamide adenine dinucleotide phosphate, reduced form (NADPH)
β-Nicotinamide adenine dinucleotide, reduced form (NADH)
Nitro blue tetrazolium (NBT)

Oil red O
Orange G

Paraformaldehyde
Pararosanilin hydrochloride
Periodic acid
Phenol
Phosphotungstic acid
Potassium aluminum sulfate (alum)
Potassium chloride
Potassium iodide
Potassium phosphate, monobasic (KH_2PO_4)
Propylene glycol

Sodium acetate
Sodium barbital
Sodium barbiturate
Sodium chloride ($NaCl$)
Sodium citrate
Sodium fluoride (NaF)
Sodium hydroxide ($NaOH$)
Sodium metabisulfite ($Na_2S_2O_5$)
Sodium α-naphthyl acid phosphate
Sodium nitrite ($NaNO_2$)
Sodium phosphate, dibasic (Na_2HPO_4)
Sodium phosphate, monobasic (NaH_2PO_4)
Sodium succinate
Strontium chloride ($SrCl_2$)
Sudan black B

Theophylline
Toluidine blue
Tris(hydroxymethyl)aminomethane (THAM) base

Wright's stain

Xylene

INDEX